WHEN GOD ISN'T GREEN

WHEN GOD ISN'T GREEN

A World-Wide Journey
to Places Where Religious Practice
and Environmentalism Collide

Jay Wexler

BEACON PRESS
BOSTON

BEACON PRESS
Boston, Massachusetts
www.beacon.org

Beacon Press books
are published under the auspices of
the Unitarian Universalist Association of Congregations.

19 18 17 16   8 7 6 5 4 3 2 1

A description of the author's visit to the National Eagle Repository, both in the
introduction and in chapter 4, was previously published, in somewhat different
form, as "Eagle Party," *Green Bag* 14, no. 2 (winter 2011): 181–89.

Library of Congress Control Number 2015018605
CIP data available from the Library of Congress.
ISBN: 978-0-8070-0192-9 (paperback : alk paper)
ISBN: 978-0-8070-0193-6 (electronic)

This book is printed on acid-free paper that meets the uncoated paper
ANSI/NISO specifications for permanence as revised in 1992.

Text design and composition by Kim Arney

For Karen

CONTENTS

INTRODUCTION

The Rocky Mountain Arsenal National Wildlife Refuge is a straight shot up Havana Street, off I-70 just east of downtown Denver, past an Office Depot and the national headquarters of a company called Scott's Liquid Gold. No signs point to the refuge, which was created on the site of a chemical munitions facility back in the mid-1990s and is now home to a herd of bison, dozens of burrowing owls, and countless furry prairie dogs. The entrance is hardly inviting, although the officer working the booth there kindly directed me two miles north to the collection of administration buildings, where I found the National Eagle Repository, a macabre little division of the US Fish and Wildlife Service that collects dead bald eagles and golden eagles and sends them (and their parts) to members of federally recognized Native American tribes that use them in religious rituals.

Applying to the repository is the primary way to legally get hold of any part of either eagle species in the United States. The Bald and Golden Eagle Protection Act of 1940 punishes unauthorized possession of eagle parts with a hefty fine and possible prison time. Although the repository serves an important need, it is highly controversial. After all, it's a rare Native American who thrills at the idea of applying to the United States government for permission to own what he or she believes is a sacred object central to a religious practice.

I had come to the repository to speak with its director, Bernadette Atencio. When I met her, she was dressed in a khaki FWS uniform, holding a mug that said CHICAGO, and looking serious. She did not seem very happy to see me, which I guess was not much of a shock—from her perspective, what good could possibly come from talking to a nerdy East Coast academic? As it turned out, however, Atencio was quite forthcoming. First we talked numbers. The repository, which has been in the area since 1995—previously it was in Oregon, and before that, in Pocatello, Idaho—receives about two thousand eagles a year, two-thirds of them bald eagles. Atencio quickly compared that number with the much larger number of pending applications—over six thousand still waiting to be filled, most of them for whole birds, rather than feathers or other parts. The eagles come from all over, generally from state fish and wildlife officials who either find the birds or are contacted by private individuals who have found them. Demand has increased significantly in recent years as the word has gotten out that the repository is the place to go for legal eagles. Large feather orders, in particular, have increased over the past few years, but these are difficult to fill as well. "Plucking takes time," Atencio pointed out. "You can't always get the perfect feathers."

Atencio talked a lot about the condition of the birds that arrive at the repository. Most have been lying out in the wild before someone finds them and sends them in. By the time the birds show up in Denver, they're often infested with bugs or in pieces. "A wing hanging here," Atencio said, "a leg hanging there." When the repository cannot use the whole bird, the staff members "piece them out," meaning that they put together the claws from one bird with the wings of another and a torso from yet another, to make a whole bird. Atencio stressed the care and detail that the repository staff exercises when processing the birds. Several times, she described her mission as basically "customer service." Running a dead-eagle processing agency is not a job for everyone, but for Atencio, who describes her work as a passion, it seems a perfect fit.

I asked if we could take a look around, and Atencio led me on a tour of the facility. From a corridor, I peered through a glass window into the cavernous, science-lab-looking room where they process the eagles. An enormous pallet stacked with nearly 170 identical

Federal Express boxes was waiting to be taken away. Each box, according to Atencio, contained a set of loose feathers. As a fact sheet explains, the repository fills orders for both "Quality Loose Feathers" and "Miscellaneous Eagle Feathers." The latter consists of "various size feathers (such as primaries, secondaries, tail, and plumes)." There is no guarantee these feathers will be any good. As the sheet puts it: "Quality may vary."

In addition to the eagles, the building that houses the eagle repository also contains the National Wildlife Repository, where the federal government stores illegally killed, traded, and shipped animals and their parts. The repository holds this material until it is no longer needed as evidence in ongoing trials and until it can be sent to a school, a museum, or another institution. We took a look at the warehouse where all the animals and their parts are kept. Seemingly endless rows of gigantic blue shelves stretched back the entire length of a space big enough to park a midsized jet. The aisles were arranged by creature, or type of creature. One nearby was labeled the "Elephant, Rhinoceros, Yak, Ostrich, Zebra" aisle. As I was talking to Atencio about something or other, I turned my head toward a different aisle. My gaze was returned by a line of tiger heads, each staring out of plastic bags right at me through their dead eyes. Several leopard heads were next to the tigers. Since I had my camera with me, I asked if I could take some pictures. Atencio said no; for security reasons, photography is not allowed in the warehouse. But to my surprise, she said we could go into the eagle processing room, and I could take pictures there. Perhaps, she added, some eagles might have arrived that needed processing. I could watch that. And take pictures.

"Really?" I asked.

"Really," Atencio said.

Sweet!

Perhaps at this point you might be wondering: Why would anybody in his right mind want to visit the National Eagle Repository? It's a good question. The answer is complicated, and it gets to why I wanted to write this book. So let me explain.

At Boston University, I've taught church-state law for well over a decade. Back in 2007, during my first sabbatical, I spent six months traveling around the country to the cities and towns where landmark Supreme Court law-and-religion cases started. I wanted to see the places and meet the people involved in these controversies that I had previously known only from law books. It was a terrific experience that resulted in my first book, *Holy Hullabaloos: A Road Trip to the Battlegrounds of the Church/State Wars*. Researching and writing that book was so much fun that when it was over, I found myself wanting to make more trips to legally significant places so I could write about them. Soon after *Holy Hullabaloos* was published, I happened to come across some cases where Native Americans claimed that they had a First Amendment right to take and possess bald eagles and their feathers. As someone who also teaches environmental law, I found these cases particularly fascinating because they pit two incredibly important interests—religious freedom and environmental protection—against each other. When I learned about the repository, which had to be one of the strangest places I'd ever heard of, I couldn't resist. As soon as I found some free time, I booked my flight to Denver.

When I returned home from the repository, I wrote a short piece about the place for a legal journal and decided I would write a book about bald eagles. Should Native American tribes be allowed to take a small number of eagles for their religious practices every year? I didn't know the answer, but I figured that by writing a book on the subject, I could work through for myself just how the government and the legal system should go about balancing the religious freedom and environmental protection interests at stake. I planned to call the book *Illegal Eagles*. The cover would have a picture of a discombobulated bald eagle, shaggy feathered and delirious. Energized, I cleared a space on my bookshelves for the inevitable Pulitzer Prize that soon would be coming my way.

As I thought more about the topic, however, I started to wonder whether the eagle controversy was an example of something bigger. Were there other circumstances involving religious practices that somehow harm the environment? I did some research and quickly learned that, all around the globe, people of faith engage in religious practices that are environmentally unfriendly. Each of these situations raises some variation of the religious-freedom-versus-environmental-protection conflict that I found so fascinating in the bald eagle context. In Colombia, demand for the palm fronds used in Palm Sunday celebrations almost rendered extinct a rare and beautiful parrot. In Israel, bonfires to mark the holiday of Lag B'Omer fill the country's air with dangerous smoke. In South Africa and Zimbabwe, members of the Shembe religion—a mixture of Christian and Zulu traditions—drape themselves in leopard pelts during spiritual dances; the demand for pelts is having a serious effect on wild leopard populations. In the United States, members of the Santeria faith sacrifice animals and splash mercury in their inner-city apartments as part of their rituals. Hindus in India throw half-burnt corpses into the Ganges River, while in Bangladesh, Hindus celebrate a Festival of Light by sacrificing tens of thousands of turtles every year. And that is just the beginning.

That's how I came to write this book. I decided that, as I did with my first book, I would travel around the country—scratch that, the *world*—to investigate in depth a few select instances of religious practices that harm the environment. My goal was to understand how this conflict has manifested itself in different situations, how various societies have tried to deal with these conflicts, and what lessons we can learn from these experiences. Nobody had ever before focused comprehensively on the relationship between religious practice and the environment, and I thought that with environmental protection and religious freedom being such important and timely subjects, this would be a great time to research and write about their interaction.

So that's what I did. For about two years, I took whatever opportunities I could to travel to some hot spots around the world and learn what I could. I went to India and Mexico, Singapore and Gua-temala, Alaska and Oklahoma, Hong Kong and Taiwan. I spoke to a Taoist priest, a Mexican palm frond collector, an Inupiat whaler, and a Taiwanese animal rights activist. I interviewed government officials in several countries to see what they planned to do about the religious practices that were causing harm to the environment, and I talked to religious leaders to see what they felt about the gov-ernment's getting involved. I hiked through palm forests high in the Sierra Madres in southern Mexico, spent a day chanting Buddhist scripture at a hilltop temple outside Taipei, and stood on a beach in Mumbai while Hindu worshippers carried twenty-five-foot idols of an elephant god into the sea. I ate way more whale blubber than I had ever expected to consume.

The relationship between religion and the environment has always been an extremely complicated one. On the one hand, many be-lieve that organized religion, particularly Christianity, has histori-cally been unfriendly toward the environment. In his classic 1966 lecture before the American Association for the Advancement of Science, titled "The Historical Roots of Our Ecological Crisis," historian Lynn White Jr. surely spoke for many when he laid the blame for much of the world's environmental problems at the feet of Christianity and its teaching that humans have dominion over nature: "Christianity, in absolute contrast to ancient paganism and Asia's religions, not only established a dualism of man and nature but also insisted that it is God's will that man exploit nature for his proper ends." Echoes of this attitude toward nature abound to-day: in the Religious Right's skepticism of science; in an Alabama official's recent claim that EPA's coal regulations interfered with a "gift from God"; in the anti-environmentalist charges of Resisting the Green Dragon, a DVD series put out by a group of conservative Christians who deny the importance of climate change and decry the environmental movement as "without doubt one of the greatest threats to society and the church today."

On the other hand, religion has done a lot of good for the environment, particularly lately. For one thing, many Eastern religions, particularly Buddhism and Taoism, as well as other traditions that treat natural objects as sacred (Shintoism and Native American religions, for example) are largely environment-friendly. Second, contrary to claims that Christianity teaches humans to act in *dominion* over the environment, many Christian believers believe that the Bible requires human beings to act as *stewards* of nature. For example, former pastor and presidential candidate Mike Huckabee has publicly cited his biblical beliefs as support for his environmental view: "We should see to it that our care for the environment enhances not only its aesthetic value but preserves the resources themselves for future generations." Finally, over the past three decades, several organized religious groups have begun enthusiastically promoting environmental values. Acting under the umbrella of a secular organization known as the Alliance of Religions and Conservation, for example, a dozen faiths from around the world have worked for many years on projects, including spreading environmental education in rural areas, promoting environmentally friendly pilgrimages, and planting trees all over Africa. In the United States, some of the loudest voices clamoring for action on climate change have come from the Christian evangelical community, many members of which have signed formal commitments to fight global warming. Indeed, the narrative of recent years when it comes to organized religion and the environment is one of groups recognizing the importance of protecting the planet and doing something about it. Pope Francis's June 2015 encyclical on the environment should put to rest any doubt about this current trend.

And yet, on the *other* other hand, as this book will show, even as religion begins to embrace environmental values, we find instances throughout the world where particular religious practices happen to harm the environment. Investigating *those* instances, and trying to figure out what to do about *them*, is the aim of this book. The more general questions about whether religion has been good or bad for the environment or whether, for example, religion has influenced major swaths of the Republican Party to take anti-environmental policy positions are fascinating and important but not the topic of this book. Here I will be solely concerned with exploring situations where the religious practice of some faith community harms the environment in some concrete way.

Perhaps you've read somewhere that I am a liberal atheist. *Aha!* you say, so the aim of this book must be to point out how religion is harmful and to explain why people should throw off their superstitious beliefs in the divine and embrace secular humanism! Nothing could be further from the truth. It's true that I'm an atheist, but I'm not one of those "new atheists" like Richard Dawkins, who thinks that religion is terrible and believes that science will save the world. I'm more of a "sad atheist." I feel the same way about God that I feel about unicorns—I think that life would be much better if they existed, and I wish I could believe that they exist, but I don't. Despite my lack of belief, I've always been fascinated by religion. I studied it as a graduate student and have written about it for two decades. I believe that religious freedom is one of the most important features of our constitutional democracy, and it drives me crazy when my fellow liberals occasionally trivialize either religious belief or religious freedom. Yet, at the same time, I really, really (really!) want a clean environment, one with fresh air and clean water and unspoiled lands and all kinds of wonderful creatures running and flying and swimming happily in their pristine habitats.

I would hope that most people share my commitment to both religious freedom and a healthy environment, and that's why I think the subject of this book is critically important. Although religion in the twenty-first century is increasingly supporting the cause of environmental protection, what happens when religious practice doesn't? What happens when a group of sincere religious believers insists on engaging in a practice that harms the environment in some way? How should we, as a society, mediate such a troubling conflict? Do we give religious freedom a trump card and let the environment suffer, to the detriment of others who will have to live with the harm? Do we instead insist that religious believers stop engaging in time-honored practices that are central to their sense of identity as individuals and as members of their communities? Or do we do something in between? And if so, what? These are the questions that animated me as I researched the book.

As I went from country to country, from conflict to conflict, I tried to understand the competing interests at stake from the perspective of those who care about them most. I tried to learn what kinds of approaches to mediating religion versus environment practices work best in what situations and thought a lot about what role law should play, if any, in these conflicts. I learned that different approaches have advantages and disadvantages depending on the circumstances. Sometimes a heavy-handed law may be the only solution, but other times the law should be used only sparingly or not at all. As with many environmental problems, sometimes it's possible to alleviate the difficulty through the development of new technologies or by harnessing the forces of the market; other times, no such fix will work. Almost all the time, education going both ways is the best first step, and a sense of empathy on all sides makes things go more smoothly.

Following an introductory chapter that provides a broad overview of the various ways that religious practices continue to place the environment at risk, each subsequent chapter tells the tale of one of my trips, each to a different area of the world and involving a different religious tradition, and I share the lessons I learned from observing a specific religion-versus-environment conflict up close. As you'll see, the lessons I took from my travels are largely aimed at the government or, in some cases, at nongovernmental organizations (NGOs), rather than at religious individuals and institutions themselves. This is because I approached my research not as a member of any particular religious tradition or indeed any religious tradition at all, but rather as a lawyer and, more importantly, as a voting citizen of our secular nation. There are certainly books to be written about how Christians should practice their religion without harming the environment or how Buddhists should change some of their practices to protect our natural resources, but in my view, those books should be written by Christians and by Buddhists, and not by me. Still, though, by the end of the book, I hope that I will have shared not only a collection of strange and entertaining stories, but also a set of insights that can help people on both sides of the religion-environment divide navigate these conflicts whenever and wherever they arise.

So, did I really get to see Bernadette Atencio and her staff pull apart some dead bald and golden eagles? Was it gross? Did it smell weird? Read on, and I promise that somewhere along the line, you'll also find out the answers to these pressing questions.

1

RELIGIOUS PRACTICE VERSUS THE ENVIRONMENT

An Overview

A three-year-old girl is poisoned when her family moves into an apartment previously occupied by mercury-sprinkling Santeria practitioners. A well-known writer contracts giardia from the Ganges River, which is littered with corpses of Hindus seeking salvation. In Sri Lanka, religious pilgrims destroy a wildlife sanctuary as they seek a holy footprint left by the Buddha, or maybe Shiva, or Adam. And across central Africa, poachers brutally slaughter thousands of elephants for ivory, which is carved into religious icons used by Buddhists and Catholics in the Philippines, China, Thailand, and Vatican City. These are just a few of the ways that religious practice harms the environment all over the globe. Here, I offer a quick tour of the problem, focusing on the effects of religious practices on the world's air, land, water, and wildlife resources, as a way of setting the stage for the more detailed accounts in the book's remaining chapters.

AIR

What more important natural resource could there be than the air we breathe every minute of every day? And yet, air pollution remains rampant throughout the world. The World Health Organization has estimated that air pollution causes seven million deaths per year from problems such as heart disease, respiratory ailments, and cancer. Major sources of air pollution include mobile sources like cars and trucks and stationary sources like factories and power plants. Relatively minor sources range from cigarettes and hairspray to volcanoes and cow farts.

When it comes to our air, the biggest hazard posed by religion is that religious people really like burning stuff. Whether they are burning incense or firecrackers or logs or paper or pieces of cardboard put together to resemble a small house, religious believers around the globe can't seem to get enough of using fire to celebrate their traditions.

Consider Lag B'Omer. This is a relatively minor Jewish holiday that young people celebrate all over Israel by lighting enormous bonfires to commemorate the death of a famous rabbi and the end of a plague that was killing a different rabbi's students. Even though I was raised Jewish, I had never heard of the holiday until my colleague Jack Beermann told me about it. Jack, who was nice enough to hire me when he chaired the Appointments Committee at my law school fourteen years ago, despite the fact that I misspelled his name in my cover letter, visits Israel often. "When I was there on Lag B'Omer, the whole country smelled like a bonfire that night and the next day," he said to me one day when I was explaining my book project to him. "Also, my clothes smelled like a bonfire, of course, so it must require lots of extra laundering."

According to news reports, there are so many bonfires lit on Lag B'Omer that satellite images reveal a smoky haze hovering over Israel during the holiday. Scientific research has shown that visits to emergency rooms for asthma and chronic obstructive pulmonary disease (COPD) occurrences go way up because of the smoke, which is hardly surprising since the concentration of particulate matter on the evening of the holiday can spike to as much as ten times the normal level. Government officials in Israel are well aware of the problem. A study authorized by the Knesset showed that the bonfires contribute to the problem of global warming, and that body has recommended (though not required) that people refrain from lighting them. The message has not been well received in most quarters. When an influential local mayor launched a campaign to convince residents to find alternative methods of celebrating the holiday, the people became outraged. As one journalist wrote: "In an instant, the popular mayor became the local killjoy, the Grinch who was trying to steal Lag B'Omer. The local press and town Internet forum erupted with residents blasting [the mayor] for his attempt to extinguish the flames. 'Next thing you know he'll be ordering us not to light Hanukkah candles,' one angry resident wrote." In fact, Hanukkah candles do contain hazardous substances like toluene, benzene, and formaldehyde, so it wouldn't be entirely shocking if somebody did try to ban them.

Beyond bonfires, the burning of incense is a fairly long-standing and ubiquitous religious practice found in all sorts of traditions, including Christianity, Hinduism, and Buddhism. Although incense can be sweet-smelling and pleasant, it is also really dangerous. For whatever reason (the smell, the context, the different treatment by the media, the extreme irrationality of all human beings), people who would go miles out of their way to avoid breathing in the smoke from a single cigarette often have no problem hanging out for hours at a temple or church where the air is filled with billowing plumes of hazardous incense smoke.

When I was visiting Hong Kong, I spent an hour or so at the School of Public Health at the Chinese University of Hong Kong, talking to a research scientist named Kin-Fai Ho, whose work focuses on the effects of toxic air pollutants on human health. Professor Ho was part of a team of scientists who were granted rare access inside of two temples in Hong Kong so they could study the effects of incense burning on the air quality. The team found that during peak times, when incense was being burnt in high quantities, the air was far more polluted than during nonpeak times. At one of the temples, for instance, the peak carbon-monoxide level was three times the nonpeak level, and the average benzene concentration was almost eight times more than the government's recommendation for public places. When I asked Dr. Ho how incense smoke compares with cigarette smoke, he said the two were comparable with respect to particulate matter, carbon monoxide, and polycyclic hydrocarbons.

Temples and the people visiting them have several alternatives that can help reduce the risk from incense smoke. In their paper, Ho and his coauthors write that "visitors may decrease the number of incense sticks burned and period of stay at temples." In my travels, I did visit temples that tried to suggest limits on how many incense sticks people should burn. Some temples have tried to deal with this problem by extinguishing incense sticks after they have been burning for a while. Particularly in Hong Kong, I sometimes saw large buckets of water standing near places where large amounts of incense were being burnt, and every once in a while, a temple worker would grab a bunch of sticks and douse them in the water. There is one suggested possible solution, however, that Dr. Ho was not very optimistic about. So-called environmentally friendly incense, which is marketed in some places as a way of reducing the environmental and health impacts of incense burning, turns out, according to a new study that Ho was working on, to have slightly fewer particulate matter emissions but little effect on the amount of toxic pollutants emitted. On my way out of the interview, looking in that journalistic way for the bottom line, I asked Dr. Ho whether he thought incense-smoke inhalation was a problem. He looked at me and responded calmly, "Yeah, it's a big problem."

Another problem is fireworks. As someone who has always hated fireworks and would rather stay inside with my head under a pillow than endure a loud, smoky Fourth of July celebration with ten thousand people staring at the sky and going "ooooh" and "ahhh" over and over for half an hour, I find it hard to understand the appeal. But still, people love watching fireworks! Every celebration these days, from the biggest national holiday to the most insignificant home-run hit by a last-place baseball team down 14–0 in the bottom of the eighth inning, seems to be marked by a blast of colorful explosions. Religious celebrations are no exception. Chinese New Year celebrations, which for some take on a religious meaning (many believe the fireworks ward off evil spirits); the Muslim holiday of Eid, which marks the end of the Ramadan fasting period; the Hindu festival of lights known as Diwali; and many other religious holidays and festivals around the world are celebrated with the abundant lighting of firecrackers and fireworks.

Unfortunately, for those of us who need to breathe air in order to live, the smoke produced by fireworks can be quite dangerous. According to one academic paper that showed the effects of fireworks on air pollution during Diwali in India, "fireworks contain harmful chemicals such as potassium nitrate, carbon and sulphur apart from an array of chemicals such as strontium, barium, sodium, titanium, zirconium, magnesium alloys, copper and aluminum powder to create the colourful effects. On burning they release gases such as carbon monoxide and nitrogen dioxide." The study concluded that fireworks contributed to excessive ozone pollution spikes during the holiday, and that "high ozone levels combined with pollution due to fireworks might be critical for elderly people and children with heart and respiratory ailments." Another Indian expert similarly concluded, "Gaseous air pollutants along with other toxic gases emitted due to burning of firecrackers aggravates the chance of attack among asthma patients. The patients with heart disease, chronic bronchitis and low immune system are also at high risk."

The realization that fireworks significantly raise air pollution levels has led officials in Beijing to call for a reduction in the use of pyrotechnics during the Chinese New Year period, and it's one reason, among others, that Abu Dhabi police have warned Eid celebrants not to use illegal fireworks in the United Arab Emirates. Even in the United States, some critics have called for the federal government to regulate fireworks, rather than exempting them from the ambit of the Clean Air Act. The EPA has refused, claiming that "Congress did not intend to require EPA to consider air-quality violations associated with such cultural traditions in regulatory determinations."

Although most people probably conjure up images of a dark and smoggy sky when they think about air pollution, in fact *indoor* air pollution may be nearly as dangerous as outdoor pollution, particularly in developing countries where people routinely burn coal and biomass fuel for cooking and heating their homes. Indoor air pollution also provides the context for one of the most bizarre examples of a religious practice that has created environmental problems in the United States.

Mercury is an element that people generally do not want to mess with. Touching it, eating it, or, most dangerously, breathing in the vapors that it releases can be extremely dangerous, potentially causing respiratory problems and damage to the nervous system. Given the perils of inhaling mercury vapors, it might be surprising to learn that some religious believers actually sprinkle the silver liquid metal inside their homes to ward off evil spirits. The practice puts not only current residents at risk but also future ones, as mercury can remain in fabrics and carpets for up to a decade, releasing dangerous vapors the entire time.

Back in 1989, a middle school chemistry teacher in Brooklyn named Arnold Wendroff was teaching his students about the periodic table. When he asked his students if they knew what mercury was used for, he fully expected someone to mention thermometers. Instead, one of his students answered that his mother, a Santeria practitioner originally from Puerto Rico, liked to sprinkle it around their apartment to fend off witches. *Witches?* Concerned and curious, Wendroff soon became a one-man watchdog of the ritualistic use of mercury. He learned that many practitioners of Caribbean religions like Santeria, Palo, and Voodoo believe that mercury can bring good luck and keep evil spirits at bay. In large US cities with substantial populations of these believers, practitioners purchase capsules containing a small amount of liquid mercury from so-called botanicas, which are essentially stores that sell religious paraphernalia.

The practitioners then do things like sprinkle the mercury on floors, furniture, or car interiors, or mop the floor with it, or burn it in candles, or mix it with perfume, or even swallow it. Because mercury vapors are so dangerous to inhale and because the mercury remains in the environment for so long, Wendroff concluded that the ritualistic use of mercury posed a significant health hazard that the government needed to address.

Through Wendroff's efforts, the EPA became aware of the problem in the early 1990s and started considering whether to do anything about it. The agency has several statutes that it could have used to regulate the ritual use of mercury inside homes, most importantly the Toxic Substances Control Act, or TSCA, which allows the agency to take a wide variety of regulatory actions against substances that pose an unreasonable risk to the environment or public health. To look into the issue, the EPA established a task force that conducted research and interviewed interested parties. Ultimately, though, the agency decided against using the TSCA, opting instead to work together with states and municipalities to spread the word about the dangers of mercury through education and community outreach.

In the wake of the task force's decision, Wendroff continued to call for further efforts to address the indoor religious mercury problem, talking to the media, writing papers in scientific journals, and interacting with various governmental units. In 2005, he asked the Office of the Inspector General at EPA to "determine whether EPA had adequately investigated whether [indoor religious mercury] contamination poses an environmental health threat and, if so, had EPA substantively acted to address its dangers." Unsurprisingly, the OIG concluded that EPA had acted properly and recommended no further action. On the other hand, the office did release a report on its investigation "to further emphasize that the ritual use of mercury poses a health risk." This final conclusion does seem to be accurate. A 2011 article in the *New York Times*, for instance, reported on the case of a three-year-old who suffered mercury poisoning when her family moved into a Rhode Island apartment that had been the site of ritual mercury use by a former tenant many years earlier.

LAND

Clean air is great, of course, but all the fresh air in the world won't make much of a difference if we have to stand on destroyed or despoiled lands when we breathe it. Our lands—our deserts and our forests, our parks and our jungles, from the fruited plains to the amber waves of grain, the mountains to the prairies, and everything else that makes up the grand landscapes of the world—are one of humankind's most treasured resources. As with the air we breathe, though, we have not always been kind to those resources. From wetland destruction to mountaintop mining to deforestation and so many other types of harm, humans continue to threaten the natural environment and put our land-based resources in great danger. Some, for example, have estimated that we have destroyed half of the world's wetlands in the past century, and others claim that we could lose the rest of our rain forests in the next century. If we don't take action to stop and reverse this damage soon, we will end up creating a bleak world indeed.

In the United States, where nearly one-third of the land is owned by the federal government, Americans have created a complicated federal system of regulations both to protect the lands and to authorize humans to use those lands in particular ways. An array of federal agencies implements these laws. Of course, state and local governments also have a good deal of responsibility for managing these lands. Like the federal government, states have their own methods for organizing public lands within their jurisdiction. And perhaps even more important, at least for most people's day-to-day lives, local governments use various land-use laws, particularly zoning regulations. Zoning laws regulate everything, from what people may do with their property (live, sell things, farm, sell porn) to how many people may live in an apartment to how big a plot of land must surround each single-family dwelling.

Once a city or town promulgates a zoning regulation, generally the only way to get around that regulation is through a variance from the zoning authority. If the facts are in your favor, you might try arguing that the regulation violates the state law that authorizes the local zoning regulation in the first place. Of course, you can always try to change the zoning law itself through the political process. Federal law, however, will rarely be of any help. Although a handful of federal constitutional challenges to zoning regulations have succeeded in the past, for the most part, nobody is going to succeed in bringing a challenge under federal statutory or constitutional law to a local zoning law.

Except, that is, when it comes to religion. Although the Supreme Court has held that the First Amendment's so-called Religion Clauses (the government "shall make no law respecting an establishment of religion or prohibiting the free exercise thereof") do not give religious believers any special exemption from general laws that burden their religious practice, in 2000 Congress passed a law called the Religious Land Use and Institutional Persons Act (RLUIPA). The 2000 law does give religious believers special rights in two very specific contexts—inmates challenging prison regulations, and landowners challenging land-use regulations. Because of this law, no governmental unit may impose a land-use regulation that places a "substantial burden" on the "religious exercise" of any individual or institution unless the government can show that the regulation is the "least restrictive means" of furthering some compelling government interest. In the parlance of constitutional law, which RLUIPA is intended to mimic, this means that the government must pass "strict scrutiny," perhaps the most stringent test created by the Supreme Court to review the constitutionality of legislation. Similarly, without passing this strict standard, the government may not apply a land-use regulation in a way that treats a religious institution less favorably than a similarly situated nonreligious institution or in a way that "unreasonably limits" a religious institution.

Although some zoning regulations are intended to do dopey things like limit the amount of vegetables you can grow in your garden or specify how big your window shutters can be or prohibit you from painting your house pink, some regulations are indeed intended to serve important purposes, such as protecting the environment. Increasingly, localities are using land-use laws as a way of supplementing inadequate state and federal regulations to protect their natural resources. Open-space requirements, watershed protections, and landscaping requirements that protect against soil erosion are all examples of how local zoning has become an important weapon in the war against environmental degradation.

So, what should happen if some church or other religious institution were to insist that a locality's zoning regulation intended to protect the environment substantially burdened the exercise of its religion? Well, it depends on a number of things, including the specific facts of the case as well as which court is hearing the dispute. Because different courts have interpreted key terms in the statute, such as "substantial burden" and "religious exercise," differently, a lot depends on where the controversy takes place. But in at least some cases, courts have held in favor of religious institutions claiming a right under RLUIPA to be exempted from some zoning law intended to protect the environment.

The most prominent of these cases comes from the lovely town of Boulder, Colorado, where the city plows the bike lanes before the streets and the restaurant menus tell you the source of every single ingredient in your farm-to-table dish. Boulder has a comprehensive zoning system that is intended, among other things, to reduce urban sprawl and maintain open space. In areas designated as agriculture zones, facilities of a certain size must apply to a county for a special-use permit; in deciding whether to grant the permit, the county considers a number of factors, including whether the new use would result in "an over-intensive use of land or excessive depletion of natural resources." In 2004, the Rocky Mountain Christian Church applied for a special-use permit to expand its already large footprint by over a hundred thousand square feet. The county denied the permit, finding that the expansion would result in an over-intensive use of the land by, among other things, causing traffic congestion and increasing the size of the church's parking area. When the church filed suit, claiming that the decision violated RLUIPA, both the trial court and the Tenth Circuit Court of Appeals agreed.

In an article critical of these decisions, University of Houston law professor Kellen Zale has argued that they "foreshadow how RLUIPA could lead to a 'death by a thousand cuts' for environmental protection across the nation." She maintains that RLUIPA can have wide-ranging effects:

> RLUIPA allows churches to do what no other land users are permitted to do: develop their property in ways that land use laws forbid. When the land use law at issue is an environmental zoning law, the threat is particularly severe. The negative environmental impacts of any one particular land use can threaten an entire ecosystem. The efficacy of environmental zoning regulations often depends on the area as a whole being protected from the effects of development; just one overly-intensive development, such as a shopping center or a school or a church, in an otherwise undeveloped or agricultural area, starts the inevitable buildup that follows. No matter that other land users are prohibited from building in wetlands by a zoning restriction, if a church can successfully argue that the zoning restriction violates RLUIPA, the damage to the environment is done.

The controversies over RLUIPA concern the surface of the land, but religion can also affect what happens underground. The relationship between religious views on burying the dead and the environment is complicated, and it is hard to know whether religion has resulted in net ecological harm on this front. Some religious traditions prohibit cremation, for example, which might be more eco-friendly than burying a chemically embalmed body in an elaborate casket on a not-insignificant area of land, but cremation comes with its own environmental hazards, primarily in the form of air pollution. Moreover, some religions insist on simple burials without embalming or extravagant caskets (or sometimes, such as in Islam, with no casket at all), and indeed the current movement in favor of green burials has received much support from religious groups.

This debate aside, most of us can probably agree that burying thousands of bags of discarded religious books and clothing in some random spot underground is not so great for the environment. This is what happened a few years ago in the central New Jersey town of Lakewood, home to one of the largest Orthodox Jewish populations in the United States.

The dispute surrounds the treatment of so-called *shaimos*, which refers to the Orthodox Jewish law that requires believers to bury certain religious books, writings, objects, and clothing separately from regular trash when these things are worn out and will no longer be used. The set of materials treated as shaimos is quite large, and although there is disagreement about exactly what counts as shaimos, many faithful believe it includes any written materials containing God's name or a verse from the Torah, as well as any object that has been used for a *mitzvah* (good deed). Such objects include, for instance, the palm frond (*lulav*) and lemon-like fruit (*etrog*), which are featured in the Jewish holiday of Sukkot. Before the advent of mass printing, burying shaimos was not such a big deal, but these days, when printouts of Torah quotations and whatnot can be copied and distributed far and wide, the amount of shaimos that needs to be buried has gotten somewhat out of control.

To dispose of their shaimos, Orthodox Jews will typically pay a temple or a rabbi to take the material away and bury it properly, under the tenets of Jewish law. They can also visit www.shaimos.org to purchase green shaimos boxes, which they can then mail, with their shaimos, to the preprinted address on the box. The website promises that the shaimos will be "properly handled and buried under [Orthodox Union] supervision." (The website also features a funny picture of a horse carrying so much shaimos in its carriage that the horse is lifted high into the air—the caption says "Don't Wait Til You Have This Much Shaimos.")

Hopefully, whoever actually ends up burying the stuff has figured out how to do so consistent with all relevant solid waste disposal laws within the jurisdiction. Federal law regulates the disposal of certain types of hazardous wastes under a statute known as the Resource Conservation and Recovery Act, or RCRA, but it leaves the disposal of nonhazardous wastes primarily to the states. The feds only require the states to ensure that nonhazardous waste is buried in sanitary, lined landfills, rather than just any old place.

Unfortunately for the citizens of Lakewood and Jackson, New Jersey, however, a rabbi there named Chaim Abadi decided to bury his congregation's shaimos not in a lined landfill but rather just willy-nilly in the woods around the towns. A few residents learned of this practice and, in 2010, contacted the New Jersey Department of Environmental Protection, which ordered the rabbi to unearth the thousands and thousands of bags that he'd buried because, among other things, the bags had been buried near a well and some protected wetlands.

The problem with Chaim Abadi's bags has not yet been solved. Environmentalists are now worried that the ink used in the written materials could end up seeping into the water. They also worry that nobody really knows exactly what is in all of the bags. The rabbi, for his part, has objected to the state's clampdown on religious practice. Eventually, the New Jersey DEP went to court and got an order requiring the rabbi to dig up the bags. As of this writing, the bags—ten tractor-trailers full, plus thousands more, loose on the ground—remain unburied. The town of Jackson has fined the rabbi (who, along with his synagogue, has had to pay huge amounts already to unbury the shaimos), and the state has offered to bury the material and "arrange for its disposition in a respectful manner," in a somewhat nearby landfill, but so far, no agreement has been reached.

When we move outside the United States to consider which religious practices might be most threatening to the world's lands, one practice clearly emerges—religious pilgrimages. Most religions hold certain places as sacred, perhaps because the place represents the origin of the tradition or continues to be a repository of the divine presence. Every year, believers make the journey to these sacred spaces as part of their religious practice. Whether the faithful are Catholics heading for the Our Lady of Guadalupe Basilica in Mexico City or Hindus off to the Ganges River at Varanasi or Muslims on their way to Mecca or Buddhists visiting the Bodhi Tree in Bodhgaya or Sikhs traveling to the Golden Temple in Amritsar, the religious pilgrimage represents one of the most important ways that devout believers connect to their traditions and demonstrate their faith. It is impossible to know exactly how big this phenomenon is, but some observers have estimated that over a hundred million people make some sort of religious pilgrimage every year. The number may be as high as two hundred million.

A couple hundred million people on the move to a few select spots around the world can wreak havoc on the environment, causing deforestation, the destruction of plant and animal life, and pollution of all sorts. For example, one study showed that in 2010, during the annual pilgrimage to Mecca for the hajj, which every able Muslim is supposed to make at least once in his or her lifetime, pilgrims left behind a hundred million plastic water bottles. Scientists in Kashmir worry that an annual Hindu pilgrimage to a cave in the Himalayas where the mark of Shiva can allegedly be seen is threatening the fragile glacial landscape there, potentially leading to the destruction of a major water source for both India and Pakistan. According to a *National Geographic* report on the problem, "The snowcapped mountains along the trail are now black with the pollution generated from hundreds of thousands of people. To reach the cave, pilgrims walk through piles of garbage, water bottles, gas cylinders, human feces, and occasional horse carcasses." And the annual pilgrimage to the top of Sri Pada, or Adam's Peak, in Sri Lanka to visit the footprint-shaped mark that Buddhists,

Hindus, Christians, and Muslims all believe is sacred, is bringing, according to one account, "garbage, sewage, putrid smells and most of all, irreversible environmental harm to the surrounding wildlife sanctuary."

In 2009, recognizing that "pilgrims often cause some form of environmental damage" and "[constitute] a threat both to the beauty of the ambiance, as well as to the flora and fauna that form part of each heritage landscape," representatives from nine religious traditions agreed to create something called the Green Pilgrimage Network to promote environmental values in connection with religious pilgrimage. The network was officially launched in 2011 at a meeting in Assisi, Italy, with a handful of pilgrimage sites signing on; the number of sites has increased substantially over the years. Each site agrees to develop a strategic vision and action plan, consistent with the relevant faith's theology, to ensure that green values will play a part in the pilgrimage activity at the site. Working with other NGOs like the Alliance of Religion and Conservation, the network urges member sites, among other things, to recycle, save water, ban plastic bags, create green transportation options, and encourage visitors to clean up the pilgrimage route as they travel. It holds conferences, engages in education efforts, and even publishes guides for practitioners, such as the very popular "Green Guide to Hajj." Nevertheless, despite these hopeful efforts, religious pilgrimages continue to cause significant environmental damage to the earth's lands across the globe.

WATER

Mmmm, water. Fresh, clean, crisp water. Water to drink, to swim in, to boat on, to catch fish in, to admire and write poems about. What would we do without clean water? And yet, even as recently as the early 1970s, the waterways of the United States were horribly polluted. Factories discharged toxic substances into the nation's rivers. Municipalities dumped untreated sewage into the rivers. Fish were dying in record numbers. Most watersheds were unusable for swimming or other forms of recreation. Lake Erie was declared dead; the Hudson River contained bacterial levels over a hundred times greater than what is considered safe. Rivers in several major northeastern metropolises could occasionally be found on fire. When Cleveland's Cuyahoga River, containing "no visible life, not even low forms such as leeches and sludge worms," burst into flames in 1969, it became clear to many that the federal government had to step in and do something. Over Richard Nixon's veto, Congress enacted the Clean Water Act in 1972, and slowly after that, the nation's waters finally started improving. Today it's even possible to swim in some of them without contracting a rash.

For a variety of reasons, including ease of enforcement and the strength of the nation's farming lobby, the Clean Water Act distinguishes between *point-source pollution*, that is, pollution from a specific, easily identifiable source, like the end of a pipe, and *non-point-source pollution*, which generally refers to pollution from runoff, particularly from agriculture. The act thoroughly and strictly regulates point-source pollution, while leaving non-point-source pollution primarily to the states. As a result, the United States has made great strides in reducing the amount of point-source pollution getting into our waters, while agricultural runoff and other forms of non-point-source pollution continue to cause significant problems.

Realizing that something has to be done about these problems, states, localities, and to some degree the EPA have begun, in recent years, to focus more on ways to reduce non-point-source pollution. Some state and local governments, for example, have begun requiring farmers and other creators of non-point pollution (timber harvesters, construction managers, and the like) to take affirmative steps to reduce the amount of non-point-source pollution generated and to reduce the possibility that the pollution will make it into waterways. These measures include things like installing fences, moving activities back some distance from waterways, restoring eroded areas, and other so-called best management practices. Although these requirements have not always been strictly enforced, environmental officials have increasingly been investigating farms and other sites to ensure that the rules are being followed. In addition to imposing fines for violations, the government also often works with farmers and other polluters to educate them about the dangers of runoff pollution and to fund pollution-reduction projects when necessary.

But what if you're a farmer who doesn't really like working with the government? And what if you don't believe in taking government money? And what if you don't go in much for developing technological solutions to problems? Enter the plain-sect, old-order Amish of Lancaster County, Pennsylvania. Amish families own over half of the five thousand or so dairy farms in the county—the county that happens to be the worst contributor to the terrible pollution that has plagued the Chesapeake Bay for decades. According to an article in the *New York Times*, Lancaster County generates over sixty million pounds of cow manure every year. And a lot of that manure makes its way to the bay, "reducing oxygen rates, killing fish and creating a dead zone that has persisted since the 1970s." As a result of this pollution, the bay's blue crabs are apparently in such bad shape and so hungry that they have started eating each other. In 2009, a group of environmental investigators checked out two dozen or so Amish farms and found that most of them were not managing their manure properly and that, as a result, a large number of nearby wells were contaminated with *E. coli*, nitrates, or possibly both.

Environmental officials have been trying to work with these Amish farmers to implement pollution control measures—such as installing fences around the farms, building larger pits to store manure, and creating buffers between the farms and nearby rivers and streams—but the suggestions have not exactly been greeted with open arms and delighted cries of "Willkommen!" (I recognize that the Amish resistance to implementing these antipollution measures differs from the other examples in the book, which involve religious practices that affirmatively harm the environment, but it warrants inclusion here because the relationship between religious belief and environmental harm remains specific and direct.) The government officials and others who have tried to convince the

Amish to change how they manage their manure have not had an easy time of it. The Amish are famously resistant to change, skeptical of the government, and wary of taking money from others. They were generally not happy to see the EPA, especially when the government first started showing up in 2009. It certainly did not help that the agency originally lacked sensitivity to the unique situation of the Amish. According to one non-Amish farmer in the region, the EPA "came in here with guns ablazing and really tried to hammer some people hard."

When the government started taking a more cooperative approach, however, it started meeting with more success. Many Amish farmers have indeed accepted funds to modernize their operations and have implemented changes to reduce manure pollution from their livestock. Even so, such steps remain controversial within the Amish community. According to a 2011 article, one farmer who accepted money from the feds didn't want his full name printed in the paper "because he was afraid his neighbors might see the story and criticize him for taking federal money."

Cow poop is one thing, but it's human poop that has in fact led to the most bitter battles between environmental officials and the Amish. The Swartzentruber Amish are a superconservative group that make your typical Old Order Amish look like Silicon Valley tech whizzes. Not only do the Swartzentrubers not use electricity or running water in their homes, but they also eschew bicycles and Velcro (Velcro!) for being too modern. Back in 2003, a group of Swartzentrubers in Pennsylvania ran into trouble when they refused to put reflective orange triangles on their buggies as required by state law. Apparently, the bright orange color was just too orangey for them. After a prolonged legal battle, the group convinced the Pennsylvania Supreme Court to let them use gray tape instead. This is not a group, in other words, that was going to cave at the first sight of a state sewage official.

When the Pennsylvania state sewage officials started showing up on Swartzentruber properties in 2008, they found numerous violations of the state sewage codes. The tanks were too small. They were made out of the wrong materials and lacked electronic monitoring equipment. Sewage was overflowing, untreated, and being emptied onto the ground. Neighbors were worried, understandably, that their well water might become contaminated. Nobody in twenty-first-century Pennsylvania wants to develop cholera. The sewage authorities told the Swartzentrubers to modernize or else. The Swartzentrubers refused to follow the law. According to one member of the group, "They're enforcing stuff that's against our religion." A judge even sentenced a guy to jail for ninety days for failing to pay his fines. Nor are these conflicts limited only to Pennsylvania. Disputes involving sewage requirements have made it to courts in Michigan and Ohio. An Ohio municipal judge, for instance, found in favor of the state's Board of Health, which had required an Amish man to install an off-lot septic system with electricity, on the grounds that "the state's interest in preventing the discharge of untreated septic/sewage from being washed downstream in the surface waters and into the groundwater is compelling."

As for the Pennsylvania Swartzentrubers, by 2013 they were planning to leave the state for New York, where they expected life to be easier. Moving to another state to avoid sewage regulations may seem incredibly radical to most of us, but it should serve as a poignant reminder of how important religion is to some people. As an attorney for one of the sewage agencies said, "I remember going to the [Amish] house . . . and the wife came out and said, 'You're going to keep me from going to heaven. Whose fault is it going to be that I'm going to hell?'"

Religious practices have caused water pollution in lots of places around the world—anything, for example, that harms the land, like the pilgrimages I talked about earlier, will also likely threaten water supplies—but the country where this problem takes center stage is India. Google "religion" and "water pollution," and you'll see what I mean. Considering the frequency, exuberance, and number of participants in religious celebrations in India, the events are bound to cause at least some degree of water pollution. Add to this the fact that so many people in India lack access to freshwater supplies, and the problem becomes extremely pressing.

Take, for example, Holi, a festival celebrated throughout India and other parts of the world with significant Hindu populations. During Holi, in addition to setting air-polluting bonfires, celebrants mark the beginning of spring by throwing abundant quantities of colored powders and liquids at each other. I have never personally witnessed a Holi celebration, but from the pictures I've looked at online it looks truly unbelievable—really wild and super fun. Everyone is running around the streets and parks and temples throwing handfuls of bright-colored powder everywhere and tossing water balloons filled with colored water at their friends and neighbors until everybody and everything is covered with a thick dusting of brilliant yellows and reds and greens and purples. The aftermath looks like a bomb went off in a Crayola crayon factory. Everyone is happy and smiling and dancing and seemingly having the time of their lives.

Except that, of course, a lot of this stuff is dangerous. This wasn't the case back when the colors were made from natural sources like flowers and leaves and turmeric, but now that most of the colors are made from chemicals, like lead oxide and aluminum bromide, the practice has become a problem. For one thing, coming into direct contact with the chemical dyes can harm your skin and eyes and throat and lungs. More to the point, though, the chemicals also end up in rivers and lakes and other water bodies, which, in India, tend to be suffering already from a great deal of industrial, municipal, and agricultural pollution.

Scientists have recognized the problem. A recent paper, titled "Impact of 'Holi' on the Environment: A Scientific Study," by two Rajasthan scientists, for example, describes the problem: "The discharge of the toxic colors in the soil and water has a deleterious effect on the water resources, soil fertility, microorganisms living in these habitats and the ecosystem integrity on the whole. These colors are not readily degradable under natural conditions and are typically not removed from waste water by conventional waste water treatments." Likewise, a number of environmental activists and NGOs have started pushing for more eco-friendly Holi practices, including a return to natural dyes. As the scientific article concludes: "We believe that large-scale efforts to increase public awareness regarding the health hazards of harmful colors, widespread availability of safer alternatives at affordable prices, and governmental regulatory control on the production and selling of hazardous chemicals will go a long way in a safer and environment-conscious celebration of this vibrant festival."

And then, of course, there is the Ganges River. This sixteen-hundred-mile-long river that flows from the Himalayas through India and Bangladesh into the Bay of Bengal is the most revered river among Hindus. Many Hindus believe that the river is the home of the goddess Ganga, a gift from the gods, or the earthly incarnation of the gods, and that bathing, drinking, or having their ashes scattered in this sacred river will wash away their sins and bring them closer to salvation. Unfortunately, the Ganges is also one of the most polluted rivers in the world. Among other things, the river is filled with garbage; dead bodies, both animal and human; hazardous chemicals like DDT and PCBs; and fecal coliforms, which are thought to be present in concentrations thousands of times higher than the safe level. Scientists believe the water from much of the river is not even clean enough for agricultural use, much less for drinking or swimming. The pollution causes all sorts of health problems, including skin rashes, infections, parasitic diseases, birth defects, and cancer, for the four hundred million or so people who live near the river. In his recent book *Being Mortal*, the well-known surgeon and writer Atul Gawande tells the moving story of scattering his father's ashes in the Ganges. The ritual requires Gawande to drink some of the river water, and even though he takes antibiotics as a precaution against infection, he ends up contracting giardiasis.

Most of the pollution in the Ganges, of course, comes from industrial and raw sewage discharge, but a not-insignificant amount of it, particularly at certain locations and during certain times, comes from religious practices. The massive amount of ritual bathing contributes pollution to the river, especially during peak periods such as the occasional Hindu festival of Kumbh Mela, the most recent of which in 2013 brought over a hundred million people to bathe in one portion of the Ganges. During the festival that year, the biological oxygen demand level of the river, which is an indication of the organic pollution present, rose to twice the recommended level on just the very first day at the site of the mass bathing. At the point of the river where it passes through the famous holy city of Varanasi, things get particularly grisly. Here, tens of thousands of Hindus are cremated every year so their remains can be scattered in the river. The cremation rituals take place on ghats, or steps that lead down from the town to the river. Often, the cremation ceremony does not completely burn the dead bodies (and other times, people who are too poor to pay for cremation simply have their corpses left in the river), so the section of the Ganges that passes through Varanasi is littered with floating corpses and partial corpses. Still, despite the fact that one estimate places the biological oxygen demand near this area as fifteen times the safe level for bathing, and despite the fact that bumping into a rotting dead body while doing the crawl stroke is a real possibility, people continue to bathe and swim in the water.

WILDLIFE

Even if the earth were surrounded by air as fresh as a baby's first breath and covered with water as clear as a glass of icy Finnish vodka, it still wouldn't be much of a place to live without animals. From our loyal pet dogs and hedgehogs, to eco-celebrities like pandas and condors, to downright weird and ugly critters like the naked mole rat or that fish with the light coming out of its head that lives on the bottom of the ocean, animals make the world about four hundred thousand times better than it would be without them (according to my own rough calculation). Even though I would expect that most people would more or less agree with this assessment of the awesomeness of animals, human beings still somehow have managed to wipe—or help wipe—entire species off the face of the planet at an alarming rate. Exactly how much havoc we are causing, extinction-wise, is difficult to pinpoint, particularly since some species would have gone extinct without any help from us. But scientists estimate that thousands of species die off every year and that the current rate of extinction is thousands of times greater than it would be without the "benefit" of human activity. Just in the past decade, for example, we've lost species like the delightful golden toad to pollution and the glorious West African black rhinoceros to poaching. Other incredible animals, like the black-footed ferret, the Sumatran tiger, and, of course, the polar bear, teeter perilously on the brink of extinction.

Religion is hardly the primary cause of species extinction, but as with the case of the bald and golden eagles, religious practice sometimes does threaten both individual animals and even entire species. There are two broad categories of religious practice that imperil wildlife. The first is animal sacrifice, where the killing of the animal is itself the central aspect of the ritual. The second involves killing an animal or animals to use them or their parts during the ritual. The latter—exemplified by the eagle situation—tends to be more dangerous from the perspective of species destruction, but the former, frankly, isn't that great for the animals either.

Religious traditions all over the world have engaged in animal sacrifice since practically the dawn of time, and many traditions still partake in the practice today. In the United States, members of the Santeria tradition continue to sacrifice chickens and goats and turtles and other animals as a way of pleasing the religion's spirits or deities, known as orishas. Back in 1990, the Supreme Court unanimously held that a Florida town violated the First Amendment by outlawing animal cruelty in a way that made it clear the town was targeting the Santeria. When I was writing my first book, I spent a day with a group of Santeria followers in Miami, including the head of the particular church that had brought the case to the Supreme Court. My visit was quite interesting, and not only because I got to hang out in an apartment with steaming bowls of goat meat and a tall machete by the front door. The followers of Santeria continue from time to time to be the object of illegal targeting by the government, but like the Amish, they are not a group that will cave easily. They know they have the law on their side, and even though they are not particularly popular among their neighbors, who tend to frown upon mass backyard sacrifices occurring next door to them, the Santeria faithful believe very strongly in what they are doing and are not going to stop anytime soon.

Sacrificing a few goats and chickens here and there may be no more harmful to the environment than hunting deer and eating a juicy steak, but some sacrificial practices around the world do pose a wee bit more danger than what Santeria is doing in Florida. Take, for example, the folk religion known as Candomblé, which is practiced in several South American countries, primarily in Brazil. Candomblé may have as many as several million followers, though the number is hard to determine, because the religion is often practiced in secret. In this tradition, which combines aspects of various African religions that entered South America through the slave trade and some native Indian beliefs, with a touch of Catholicism thrown in for good measure, practitioners sacrifice all sorts of animals to please the spirits. Although most of the animals are nonexotic ones like pigeons, guinea fowl, and goats, one sacrificial animal is the huge yellow-footed tortoise, which is considered vulnerable to extinction by international experts on species destruction. Apparently, practitioners of Candomblé sacrifice the tortoise specifically to please one particular spirit named Shango, because like the deity, the tortoise is considered especially powerful.

Endangered turtles are also having a hard time of it in Bangladesh. In the Hindu festival of Kali Puja, celebrants demonstrate their devotion to Kali, the goddess of power. The festival occurs during the autumn and is particularly popular in Bangladesh. One way that practitioners celebrate the holiday is by killing and eating turtles, because they believe that eating the turtle meat will give them strength, like the goddess. As a result, every October or November witnesses a mass slaughter of turtles, terrapins, and tortoises in this south Asian country. During the 2011 festival, for example, perhaps as many as one hundred thousand turtles were slaughtered. You can look online if you want to see some footage from this celebration, but I wouldn't recommend it. The turtles are often unceremoniously hacked up and sold right on the street; the aftermath is a sea of blood and mountains of empty shells.

If this wasn't bad enough, the sacrificer and the eater believe that the scarcer the turtle the better. Ten percent of the world's turtles live in Bangladesh, and many of the twenty-two freshwater species of turtles and tortoises that can be found in the country are considered endangered. Bangladeshi law makes collecting these species illegal, but a large network of turtle collectors and vendors work together throughout the country to evade the law, and officials have largely looked the other way. During recent Kali Puja celebrations, observers have allegedly seen a number of endangered species being sold and eaten at the markets. These species include the northern river terrapin, one of the most endangered species of turtle on the planet, and the black softshell turtle, which until 2002 had been classified as extinct in the wild until it was rediscovered in a river in northeastern India. According to Dr. S. M. A. Rashid, an expert who has worked for many years to stop the illegal trade in endangered turtles in Bangladesh, the effect of the holiday on turtle populations in the wild was "devastating." When I asked Dr. Rashid by e-mail about the current situation in Bangladesh, he told me that in the last couple of years since 2012, media attention has forced the government to step up its enforcement efforts. Open sales of turtles stopped for some time, and some village markets that sold turtles as part of the Hindu festival were raided as well. But Dr. Rashid also said that as a result of the increased enforcement efforts, much of the turtle trade has simply gone underground. These days, turtle traders have begun "home delivery" as an alternative to selling in public; an interested buyer need only place a mobile phone call to a trader, and the trader will deliver the possibly endangered turtle to the buyer's home.

Two stories from Africa illustrate the second general type of problem posed by religion for wildlife—the use of animals as part of a religious observance or ritual. I'll start with big cats. One of the fastest-growing religious groups in South Africa—now numbering at least five million adherents—is the Nazareth Baptist Church or the Shembe, a combination of Christian and Zulu traditions named after Isaiah Shembe, a charismatic leader who began the religion in the early twentieth century. When believers get together in large numbers to worship with prayer and rhythmic dancing, they don elaborate costumes of monkey-tail loincloths, headgear made with ostrich feathers, and leopard-skin pelts worn over their chests. The leopard pelts are meant to symbolize power, and thousands of leopards, a species that is listed as "near threatened" and getting worse, have been killed for these rituals. As with the turtles in Bangladesh, sale and possession of the leopard skins is technically illegal, but officials turn the other way and fail to enforce the laws.

A few years ago, a South African biologist and conservationist named Tristan Dickerson came up with an ingenious plan to save the leopards. He decided to try developing a fake leopard pelt that would look real enough for at least the rank and file Shembe practitioners to wear instead of the real ones. At first, things did not proceed smoothly. For one thing, the Shembe followers were not particularly worried about the leopards. Many did not know that leopards were in danger of becoming endangered, and some even told Dickerson that if the number of leopards got too low, the Shembe messiah would simply make more of them. The other problem was that Dickerson couldn't get the pelts to look right. His attempt to print leopard spots onto impala coats, for instance, was not fooling anyone. Finally, though, Dickerson worked with expert fake-fur makers in China to develop a pelt that the church could get behind. In early 2014, the church endorsed the practice of wearing fake pelts, and as of early that year, nearly two thousand pelts had been given out to believers for free. One estimate put the number of fake-pelt wearers at 10 percent of the church, with another estimate predicting that 70 percent of the dancers would be wearing fake pelts within a couple of years.

Then there are the elephants. You will find no sadder story in this book than the one about the elephants. Most people know that poachers kill elephants in Africa for the animals' ivory tusks, but do you know how many elephants are actually killed? A recent study—apparently the most rigorous one to date—concluded that the previous estimate of about twenty-five thousand per year was *too low* and that in fact poachers had killed one hundred thousand African elephants over a three-year period between 2010 and 2012. *One hundred thousand elephants.* Try to imagine, if you can, a pile of one hundred thousand dead, tusk-less elephants. In a week, when you're done weeping, move on to the next sentence. Some experts believe that if the current rate of killing continues, elephants could be extinct within a decade.

Until journalist Bryan Christy published his now classic *National Geographic* article "Blood Ivory" in 2012, few knew how much of this illegal slaughter and ivory trade can be traced to religion. In that piece, Christy tells of his travels around the world to find out what all the ivory is used for. He learns that much of it is made into religious icons. In the Philippines, millions of Catholics come together in January with their idols of Santo Niño de Cebu to celebrate the country's hugely important religious figure. Because "many believe that what you invest in devotion to your own icon determines what blessings you will receive in return . . . the material of choice is elephant ivory." Christy visits ivory carvers in Thailand who produce ivory Buddhist icons for sale throughout the nation ("Ivory removes bad spirits," a monk tells him); an ivory carving factory in Beijing that "smells and sounds like . . . a vast dentist office" where workers make sculptures of Buddhas and a variety of other Chinese folk gods; and Vatican City, where ivory idols can be found for sale in the stores on St. Peter's Square. "No matter where I find ivory," Christy writes, "religion is close at hand."

In the aftermath of Christy's story, Oliver Payne, who edited the piece for *National Geographic,* tried to use Christy's findings to pressure the Vatican to take a stand against the killing of elephants for religious uses. Payne figured that if the pope (it was Benedict at the time) were to make a strong statement condemning the slaughter of elephants, it might have an important effect on the worldwide ivory trade. He sent a letter to Father Federico Lombardi, the Vatican's Press Office director, asking about the Vatican's position on using ivory for religious objects. When Payne hadn't heard back from Father Lombardi after two months, the editor sent a follow-up missive, suggesting that "the Vatican could make an important contribution to both humankind and the environment by taking a few important steps, in particular: (1) Declare the use of ivory for religious purposes as no longer acceptable; (2) Call for an immediate halt to all carving and exchange of ivory for religious and commercial purposes; (3) Accede to the Convention on International Trade in Endangered Species of Wild Fauna and Flora." And then he waited.

Finally, in January 2013, four months after Payne sent his first letter, he received a response from Father Lombardi. Although Lombardi stated that "we are absolutely convinced that the massacre of elephant is a very serious matter," most of his letter involved pointing out how the ivory trade really isn't the Vatican's fault. Philippine Catholics who are "responsible for illegal trade in ivory?" Not the Vatican's fault. The store that sells ivory devotional objects "a few dozen meters" from Lombardi's office? It's privately owned and not the Vatican's fault. The Asian countries primarily responsible for the trade in ivory religious objects? Catholics there are a "tiny minority"—it can't possibly be the Vatican's fault. All of this was true, of course, but beside the point. Payne hoped the Vatican would take a stand against the slaughter of elephants, and the Vatican responded by pointing fingers and making a hollow promise to "raise awareness of the problem through programming on Vatican Radio." Although Pope Francis, who took over for Benedict a few months after Lombardi issued his response, had already done many great things by the end of 2014, he had made no public statements about the slaughter of elephants.

Since around 2012, some progress has been made. The work of Christy, Payne, and many other dedicated conservationists in Africa and elsewhere around the world has done much to raise consciousness about the plight of the elephants. A number of countries, including the United States, the Philippines, and Hong Kong, have destroyed their stores of ivory seized from illegal traders. This is no small amount of ivory. The Philippines destroyed five tons; the United States six. Hong Kong has started destroying its ivory and has pledged to destroy nearly thirty tons over the course of a year. These are important gestures, and they help spread the word about the elephants, but at the same time it is not entirely clear how much good they are doing. The ivory being destroyed, after all, has already been removed from the market, and the poachers' knowledge that this material can't make its way back into the market may only serve to increase prices for new, illegally procured ivory. There is, in short, a lot of work left to do if we want to ensure that religion won't end up eradicating elephants from the face of the earth.

As I hope I have demonstrated, the problem of religious practices that harm the environment is a real one. It occurs all over the world, including inside the United States, and it involves a wide range of religious traditions, both large and small. The large, looming question, then, is how society, and particularly the government, should seek to protect the environment without unduly burdening the freedom of religious believers to practice their faiths. The trips I describe in the rest of the book represent my attempts to find some answers to this critical question.

2

GUATEMALAN GREENERY

International Effects and the Influence of Markets

I sat in the passenger seat of Juan Trujillo's pickup truck, careening through a dense forest in northern Guatemala. The road we were driving on—and I use the word *road* here with some hesitation—was simply a deeply grooved dirt trail that had been coarsely etched through the jungle so that trucks like this one could travel between the town of Flores, a lovely little place overrun by tourists on an island in the middle of a scenic lake, and the many tiny villages and Mayan ruins that dot the northern jungles. Rains had turned parts of the road into muddy mush, and the trip was about as smooth as the swirling teacup ride at Disneyland. I was being jarred in all directions, my head occasionally thumping up against the roof of the truck, as I tried to do some basic fourth-grade mathematical calculations using the numbers that I had written down in my little notebook at the village we'd just departed.

In that village—it's called Carmelita—as in several other villages in the northern Guatemalan region of Petén, farmers make a modest living by selling products harvested from the forest itself. These include both wood products and nonwood products like pepper, a tree gum called chicle that is used to make chewing gum, and ramón, a seed that apparently the Mayans munched on while constructing their intricate temples and that the Guatemalan chamber of commerce would like to market as the next international superfood of the twenty-first century. The farmers also tend to various types of palm trees. Not the tall, swaying, tropical, coconut-bearing palm trees of Southern California or Miami Beach, but squat little plants and bushes that sit in the shade of the forest and whose leaves are harvested and sent overseas to be used in floral arrangements and carried aloft by celebrants during the spring holiday of Palm Sunday.

Palm Sunday is a Christian holiday that falls on the Sunday before Easter and commemorates the day that Jesus Christ entered the holy city of Jerusalem while crowds of admirers laid down palm leaves on the ground before him. Like most people, before I started working on this book, I had never considered for a second where the palm leaves used every year for this holiday actually come from. It turns out they come from a handful of Latin American countries, including Guatemala, Belize, and Mexico, and it also turns out that for the most part they are harvested in a haphazard fashion without any regard to the long-term sustainability of the forests. There are, however, some exceptions. In a few select villages in northern Guatemala and the southern Mexican state of Chiapas, local farmers—with the help of savvy non-governmental environmental organizations, forward-looking governmental agencies, and a conglomeration of religious organizations in the United States that have come together under the leadership of an unassuming University of Minnesota professor to form the so-called "Eco Palm" project— have revolutionized the way that palm leaves are harvested in the region. In February 2012, I spent a week in Guatemala and Mexico to learn about these harvesting techniques so I could understand what role religious groups have played in both creating and solving the problem of palm deforestation. But why was I furiously doing long division in the front seat of a pickup truck tromping through the Guatemalan jungle? We'll get to that.

I first learned about the potentially deleterious effects of Palm Sunday on Latin American forests when a colleague pointed me to a fascinating article about a Colombian bird published in *Audubon* magazine. In the piece, a terrific freelance writer named Susan McGrath tells the story of the yellow-eared parrot, a glorious creature, mostly bright green with yellow patches on the side of its head, that lives in a small area in the Andes Mountains of Colombia. The yellow-eared parrot is highly sensitive and can only nest in a single type of wax palm tree. Unfortunately for these feathered fellows, however, the local church in the region where the birds live had been using this tree's leaves for its annual Palm Sunday celebration, thus decimating the tree population and driving the parrot to the brink of extinction. When the bird's defenders first tried to convince the town's priest to use a different type of palm, they were not well received. Specifically, according to McGrath, the priest "blew a gasket" and "admonished parishioners to stand fast and keep using the palms." It was only when the priest was transferred and replaced by a more flexible religious leader that the bird lovers started to make progress. In the end, the church replaced the wax palm leaves with an abundant native species called *iraca* (which some still oppose because it is too "puny"), and as a result the birds have made a remarkable recovery.

The Eco Palm project that I mentioned earlier is another super-awesome effort to protect the unique environment where Palm Sunday palms grow. It can circuitously be traced back to the North American Free Trade Agreement (NAFTA), the 1994 accord between the United States, Canada, and Mexico that eliminated tariffs and other impediments to free trade among the three countries. One argument that free trade opponents often raise is that lifting barriers to trade can sometimes harm the environment. An importing country that cares a lot about the environment, for example, cannot always insist that an exporting country do things in an environmentally friendly way as a condition on importing the exporter's goods.

An example of this problem occurred in the mid-1990s, when the United States insisted that it would only import shrimp caught in nets with a turtle-excluding device that would prevent shrimpers from inadvertently ensnaring and killing endangered sea turtles. This requirement was fine for big, rich Western shrimpers, but for poor independent shrimpers in places like Malaysia and Pakistan, installing these turtle excluding devices was completely unfeasible (the shrimpers' yearly income in some cases was about the same amount as the cost of one device). These small shrimpers brought a high-profile case against the United States in front of the World Trade Organization, which ultimately held for the United States, but not without placing some limits on our ability to insist on environmentally friendly processes as a precondition to importing foreign goods. Some environmentalists were really angry about this. If you remember back to 1999, when a loose and weird coalition of unions and environmental groups and anarchists (anarchists!) protested the WTO's Seattle conference and caused all sorts of property damage—some called these events the Battle of Seattle—you might recall that a bunch of the protestors were dressed up as turtles. This is why.

Anyway, due to these concerns about free trade, the three NAFTA countries established an organization called the North American Commission for Environmental Cooperation (CEC) to study NAFTA's environmental effects and to take steps to mitigate them. Sometime in the late 1990s, the head of the Trade and Environment Program at CEC, a woman named Chantal Line-Carpentier, became aware that palm forests in southern Mexico were being harmed by unsustainable harvesting techniques. Line-Carpentier knew that consumers in the United States are often willing to pay a premium for products they know are helping (or at least not hurting) the environment. She had previously worked on projects to harness private market demand as a way of promoting sustainably grown coffee, and she wondered whether the same approach might help the palms. In 2000, she contacted Dean Current, a forestry economist at the University of Minnesota who had a great deal of experience with Latin American forests, and asked him if he would conduct a study to figure out whether such a thing might be possible.

At first, Current was skeptical about doing the study. As he told me when we met in his office in the University of Minnesota's Soil Science Building in St. Paul, he is trained as an economist, not a marketer. But he eventually agreed, and along with a graduate student, he not only visited and talked to all the wholesale florists in the Twin Cities but also did a nationwide study of the entire palm importing business in the United States. He learned that although most of the demand for palms from Latin America comes from florists who use the palms as part of floral displays, churches that buy palms for Palm Sunday also make up a significant portion of the market. Plus, many of these churches seemed to be quite willing to pay a little extra for palms that were sustainably harvested, if the churches were guaranteed that the additional money would go back to the communities to support sustainable practices and other important community needs, such as education. As Current put it when I asked him if any churches refused to participate in his program: "No one says that they don't want to pay twenty more dollars for sustainability." And thus, the EcoPalm project was born.

The project started small. In 2005, its pilot year, Current and a graduate student took orders themselves from religious congregations in Minnesota and North Dakota and delivered a total of about five thousand palm fronds to the churches from the back of a van. Over time, however, the project grew substantially. Current brought in religious organizations like Lutheran World Relief and the Episcopal Church to spread the word to their congregations, and he had a professional florist take care of the sales and deliveries. Media stories, like one published in the *New York Times* in 2007, brought the project added publicity. By 2012, the project filled orders for over nine hundred thousand fronds. This was still a relatively small percentage of the palms that are used in Palm Sunday services around the country, but it marked a great advance over the situation just a few years earlier, and it meant that a good deal of money was being sent to communities in need.

It was clear from what I had read about the EcoPalm project and talking to Professor Current that I had happened upon a very cool initiative that was providing real benefits to the environment. But I still had a lot of questions. What were these harvesting techniques that the villagers were using to protect the palm forests? What do palm forests even look like? Do the villagers know who was buying their palms, and why? Most importantly, how did these select villages in Guatemala and Mexico happen to develop sustainable harvesting techniques? Did the EcoPalm project, with its promise of paying extra for sustainably harvested palms, create the incentive to start harvesting the palms in a sustainable fashion, or were the palms already being harvested sustainably in some places? Reading the newspaper articles about the project, I believed that the EcoPalm project itself had created the incentives. The *New York Times* piece, for instance, suggested that the primary consumers of the sustainably harvested palms were indeed the churches. Was religion really that important of a factor here, either in creating or solving the palm forest crisis in Latin America? I wanted to learn more. And so I traveled south.

In 2013, I left for Guatemala in mid-February, which is a good time to leave for Guatemala if you live in Boston, where just the week before, we had received twenty-three inches of snow during a single storm. I had never been to Central America, so I was a little worried about the things that people who have never actually gone to Central America tend to be worried about when they go to Central America, like whether it was safe. The assistant dean who approved my funding for the trip didn't exactly put me at ease when he told me there was "no money in the ransom budget, so be careful." As it turned out, though, I had no use for such funds. The trip went as smooth as can be.

One thing that made my research in Guatemala and Mexico more difficult than it needed to be was my poor language skills. My Spanish is *muy crappy*. I should have taken Spanish in high school, but instead I took Latin, I guess because I thought it was more important to be able to talk with the pope than with the 350 million or so people who speak Spanish all over the world. Anyway, a couple of years before I took this trip, for reasons I won't get into, I decided to start studying Spanish. My university has a great perk that allows faculty members to take courses in other parts of the school for free. So one summer, I enrolled in Introduction to Spanish, along with fifteen nineteen-year-olds. Let me tell you, when you're an old guy and you take a class with a bunch of kids who were born after you graduated from college and who are constantly playing with their phones while the teacher is talking and who drink twenty-four-ounce Red Bulls at nine in the morning, it's really weird. I kept expecting them to ask me to buy them beer (they never did). Well, long story short, I ended up taking three semesters of Spanish this way, and although I certainly did learn something (witness the magnificent and heart-warming semester-ending play in which two of my teenaged classmates and I reimagined the ending to the hit biopic *Selena*), my Spanish still leaves a whole lot to be desired.

After about twelve hours of flying and sitting around at the Guatemala City airport, my turboprop airplane touched down at the tiny Flores airport, where I was met by Juan Trujillo, a squat, barrel-chested guy who works for a great NGO called Rainforest Alliance, where his official title is non-timber forest products coordinator. Juan is originally from the village of Carmelita (the place we were coming back from when I was trying to do my math), and his career has taken him back and forth from running the nontimber products cooperative in the village to working in the city for the NGO. While I was in Guatemala, Juan worked twelve-hour days and drove his truck for what seemed like thousands of miles to help me understand what was going on in his country. Because Juan's English is about as good as my Spanish, however, he brought along Celeste, a young woman with excellent English whose prior translating job was with a United Nations soldier force in Haiti. There was some trouble getting me to my hotel, as a religious procession for some saint had closed down some of the roads on the island, but before

long, I was checked into a decent if a bit grungy place called Hotel Casazul, where I practiced my terrible Spanish with the lady in
charge to try to figure out how to access the Wi-Fi system (I failed).

The schedule for my three days in Petén was jam-packed with travel
to isolated villages and meetings with NGO people, government
representatives, and others involved in the palm trade. Over the
course of these busy days, I tried to figure out what was really going on with the palms in Guatemala the best I could, but it was not
always easy. Part of the problem was the language—even though
my translator was great, speaking (and listening) through her was
not the same as speaking directly to the people who were explaining things, and much meaning inevitably got lost. I also came with
various preconceptions from my reading and just my general background, and some of these preconceptions were difficult to shake.
I gained some key understanding early on, fortunately, over breakfast by the lake with Juan Trujillo and José Román Carrera, a bigwig
at the Rainforest Alliance for Central America and the Caribbean.
I had simply assumed that the villages I was about to visit were all
on privately owned land. In fact, as Carrera explained to me, the
villages are all located on land that is owned by the government.

Specifically, the villages are all within the Mayan Biosphere Reserve, a 21,000-square-kilometer government-owned area that was
set aside in 1990 to protect the extremely valuable forest resources
of northern Guatemala. A government agency known as CONAP
manages the area, which is split up into three types of land desig-
nations—*core* zones, the primary ecological sites where nobody is
supposed to do anything that would harm the forests; *buffer* zones,
where some farming is allowed; and *multiple-use* zones, where the
government may enter into concession agreements with communities to do certain types of more invasive activities, including harvesting of timber and nontimber products like the palms.

Beginning in the mid-2000s, CONAP entered into concession agreements with at least a dozen community organizations within the multiple-use zones to allow the communities, under very specific terms and with extensive monitoring, to harvest and sell various wood and nonwood products. The arrangement has been incredibly successful. According to Carrera, the rate of deforestation in the community areas is twenty times less than in the rest of the reserve, even than in the core zones, which are supposed to be pristine. This last part took me a while to understand, I have to admit. In the United States, a protected zone like a wilderness area or a national monument is almost guaranteed to be in pretty good shape, ecologically. But in northern Guatemala, the supposedly unspoiled lands are often destroyed by narco-traffickers, who clear huge swaths of land out of the forest and who the government is either unable or unwilling to stop. In the communities, however, where the villagers know that the health of the forest is critical to their own future (and where the government and the NGOs provide them with a lot of help), the forests remain more or less protected.

By this time, I was incredibly eager to actually visit one of these villages, so as soon as the first of about fifteen meals of eggs and tortillas and beans I would consume on the trip was finished, we set off for the village of Uaxactun (pronounced kind of like "Washington"). A community of about a thousand people approximately ninety kilometers north of Flores, Uaxactun is located near a set of impressive Mayan ruins (for what it's worth, the sequel to the original Activision videogame hit *Pitfall*, known as *Pitfall: The Mayan Adventure*, was apparently based here). Our group included Juan, Celeste, Jorge Sosa (one of the greatest guys ever), and me. A thirty-year forestry veteran who works for the association of forest communities called ACOFOP (Asociación de Comunidades Forestales de Petén), Sosa knows the name of every tree, seed, and leaf in the forest and wears a white sombrero at all times. Everyone loves Jorge Sosa, even though they all give him grief about the hat. With his dark features and slight frame and that hat, he looks like somebody out of an old movie, so I surreptitiously tried to take as many pictures of him as I could. I even drew a little sketch of him in my notebook and still entertain plans of painting a portrait of him someday if I ever start painting again.

The drive to Uaxactun took about three hours, during which time I kept my eyes peeled for jaguars and monkeys and other exotic Guatemalan fauna frolicking in the forest. Sadly, I did not see so much as an interesting spider during my entire trip. We arrived in the village around midday. It was a charming place. The main living area consisted of a series of little houses surrounding a cleared-out rectangular area maybe three football fields long, which had been used as an airstrip at one time. Pigs and horses and dogs ran freely around; kids were playing in the field and looking at me with my pen and little notebook as if I were a weirdo.

I was escorted into a small wooden house, painted green with a tin roof and concrete floor. There I met with four community leaders who explained how palm harvesting had changed in the village over the past ten years. There were three men and a woman. One of the men looked remarkably like a darker version of Burt Reynolds in *Cannonball Run*, and this distracted me several times during our conversation because I kept expecting Dom DeLuise to show up. The four villagers explained to me how before they entered into a concession agreement with CONAP in the mid-2000s, the *xateros* (in Guatemala, the palms are called *xate* and the palm-cutters are called *xateros*) would cut as many palms as they could as fast as they could, because the buyers paid for the palms solely by quantity. The buyers would pay the xateros very little and then send the palms on to Guatemala City, where an exporter would sell them to overseas distributors. As a result, the xateros were destroying the forests.

Beginning in 2005, however, the village, along with government and NGO forest engineers, developed a management plan that totally changed how the palms were harvested. Now, the community's forest is divided into four zones. Harvesters are allowed to cut palms in only one of the four zones at a time and only for thirty days, after which they must move to a different zone. Consequently, the palms in each zone get ninety days to recuperate before harvesters return to cut some more. Also, the xateros were trained to cut the palm leaves in a way that benefits the palms. Instead of cutting all of the plant's leaves, regardless of their quality, the xateros now only cut the leaves that are good enough to be sold overseas. Leaves that have holes, or that are yellowed, or that are otherwise flawed are left alone; they may not be perfect enough to be put into a floral display or carried on Palm Sunday, but they are still very important for the health of the plant and thus the forest and the entire ecosystem.

The community sells the palms directly to an importer in the United States (how this came about is an important part of the story, one that I did not learn about until later on in the week, so stay tuned), and to ensure that the xateros are only cutting the good leaves from the plants, the harvesters must bring the bags of leaves to the *bodega de xate*, a special building where village women sort through the leaves, note how many of the leaves received from each xatero are good ones, and decide how much each xatero will get paid for his harvest. The part about the women being the ones who run the sorting part of the operation is universal in these sustainably harvesting communities, and it is one of the great benefits of the new approach to palm harvesting. Before these developments, women in the village were often unemployed, but now they participate in the work of the community while also bringing home extra money to help their families.

Once I got the basic gist of what was going on, I asked about the contribution of the US churches to the village. After all, although I was learning a lot of fascinating stuff about internal Guatemalan conservation efforts, my original goal in studying these changes was to understand whether and how religious groups in the United States have contributed both to the problems and to the solutions here in Latin America. The villagers were extremely positive about their relationship with the EcoPalm project. According to the four people I was meeting with, the churches had given the community a great deal of extra money above and beyond the normal price of the palms. (I had not yet gotten into exact amounts—this was what would end up getting me all flummoxed and doing long division on the way home from Carmelita.) The community has used the money for all sorts of things—not only to keep up its forest management practices and to get those practices certified by Rainforest Alliance (which the government requires), but also for infrastructure and education. In the past, for instance, the villagers used the money to hire teachers for older students and to put a concrete floor into a school so the children wouldn't have to study on dirt.

Our next stop was the bodega de xate, a long, open warehouse constructed from wooden slats painted turquoise and topped with a tin roof. On the side of the bodega, a group of students had painted a mural depicting the steps of the palm harvesting and selling process—from a xatero cutting the palms in the forest, to a woman selecting the usable palms in the warehouse, to a truck transporting the palms to a refrigerated room in Guatemala City, where they are stored before being sent overseas. Inside the bodega, a dozen wooden tables were covered with palms, some in rough piles, unsorted, just in from the forest, and others in neat stacks of uniform bundles of twenty palms each. Burt Reynolds and some of his friends showed me how they wrap thirty bundles together in brown paper to make a package of six hundred fronds—the basic unit in which the palms are exported. An older woman wearing a short blue skirt and a black top explained how she sorts through the just-harvested palms and throws the ones that are too small or misshapen or otherwise flawed on the ground. The woman's name was Raina Isabel Valenzuela, and she showed me a notebook in which she keeps careful notes about how many bad palms each xatero brings in. She claimed that before the harvesters instituted the new sorting system, nearly 85 percent of the harvested palms were unusable; now, she said, very few are discarded. If the records show that a xatero has been bringing in too many unusable leaves, then the villagers will have a talk with him, because every unusable leaf that is cut from a tree harms the forest without getting anything for the village in return.

Although it would take until the Mexico part of my trip for me to really hike into a forest to see these palms in their natural habitat, we did take a walk to the edge of the forest in Uaxactun, where Jorge Sosa and the guys showed me the different kinds of palms that are grown there. Only the so-called *xate jade*—a frond of about eighteen to twenty-four inches with maybe twelve or fourteen leaves on each frond—are really exported in serious numbers. There are a couple of other types as well, including one called fish tail (*cola de piscado*), which looks like the tail of a fish, and a narrower palm called the *xate hembra*. The narrower palm used to sell but now, much to the chagrin of the villagers, has become nearly impossible to export. (At one point, the villagers pulled me over and implored me to convince more of my fellow Americans to import this type of palm—which is what I'm doing here, now, in this sentence.) The villagers also gave me a little primer on how they cut the palms to keep the forest healthy (don't cut the principal stem, only cut leaves of a certain size) before we had to say our farewells and hit the so-called road back to Flores.

Although my days were filled with activities planned by Juan and the other good folks at Rainforest Alliance, I did get a little time at night to relax by myself. My hotel room had a nice balcony overlooking the lake, so I sat out there as the sun went down, reviewing my notes and drinking a can of Gallo, the national beer whose label sports a squawking rooster. Being a tourist town, Flores has a lot of really good restaurants, so that night, I went to an Italian place with a leafy terrace. I sat down on the far right of a tiny bar with only three stools. While I was drinking another Gallo and waiting for my dinner, a gray cat jumped up onto the far left stool and sat down. I said hello to the cat. The cat ignored me. Then the waiter put a shot glass with a piece of ice in front of the cat, and the cat licked it. I sipped my beer. The cat licked the ice. Together we drank in silence.

Juan and I got on the road to Carmelita the next day early enough that when we arrived, the first thing we did was eat breakfast. This is the village where Juan is from, so he knew just whose house to go to for a morning meal. We sat in someone's outdoor thatched-roof kitchen, chickens wandering around clucking, the radio playing Spanish music (including, unbelievably, "Macarena") in the background. A wood stove was heating up a metal plate on which was set a large pot of water that we dipped our cups into for instant coffee. The coffee situation was puzzling to me at first. Guatemala is one of the great coffee-producing countries in the world—indeed, when I left on my trip, I had a big bag of Peet's Guatemalan coffee in my cabinet at home—but all of the coffee I had been served since arriving was instant or otherwise not too good. Somebody explained the obvious to me later on in the trip—the people who harvest the coffee are poor, so they need to sell the best coffee to big companies that can sell the coffee to people like me in the United States, who will buy it for three dollars a cup, which means that the coffee that's left in Guatemala for the coffee harvesters to drink is fairly terrible. But the eggs and beans and tortillas that Juan's friend cooked up for us on the outdoor wood stove were definitely delicious. While we ate, Jorge and Juan chatted about their many grandchildren and how hard it can be to get kids to stop looking at some kind of screen, which is something I think about all the time when I'm back at home.

Carmelita is a bit smaller and farther out than Uaxactun, but it was otherwise quite similar. As in Uaxactun, we started by meeting the village leaders to discuss their new harvesting techniques (which are much the same as in the previous village) and what they purchased with the extra money donated by the churches (not just education, but health benefits, life insurance, fire protection, and even coffins). Then we hit the bodega de xate. It was here that I started confusing myself with numbers and math. I had realized by this second day in Guatemala that my original notion of the relationship between the EcoPalm project and the development of sustainable harvesting practices was not correct. The EcoPalm project itself was not, as I had originally assumed, the *cause* of these sustainable practices. Instead, the villagers (with help from the government and NGOs) had developed the sustainable practices independently, and then the churches in the United States basically decided to invest in the villages by agreeing to buy palms from the Guatemalans for a premium. But how much of an investment were we talking? Have the churches been a major player in supporting these villages, or did religion just have a bit part?

I asked the woman who was showing me around the bodega how many more palms the villagers sell around Palm Sunday than they sell during the rest of the year. She told me they tend to sell about two hundred extra packages during those weeks. At 600 palm fronds per package, that's an additional 120,000 palms for the holiday. Each package costs, on average, about twelve dollars. So far, so good. But then when I asked how much the churches paid as a premium, I thought she said that the churches pay an additional sustainability bonus of five cents for every dollar that they pay in total for the palms. Now, I have never been all that good in math, so I couldn't really process the numbers in my head, but I started to think that maybe we weren't talking about that much money. This concern was nagging at me while Celeste, Juan, Jorge, and I were eating lunch on the patio of one of the villagers' homes. As I nibbled at a chicken leg served in a bowl of soup with rice and potatoes and worried about whether the red Kool-Aid that I was drinking from a small plastic glass with a smiling cow on it was made with purified water, I started to wonder whether I had come all this way to Central America to research a program that was donating but a piddling amount to the sustainability of this critical ecosystem.

And so that's why I was trying to do some basic math in Juan's truck as the four of us—Juan and I up front and Jorge and Celeste in the backseat—bounced our way out of Carmelita and back toward Flores. It was a strange ride for other reasons too. As usual, dogs and turkeys and pigs were all over the road; several chickens crossed the road, seemingly, truly, just to get to the other side. At several points, Juan pulled over to pick up random people to give them a lift in the flatbed. I guess this is just the custom where we were, although I was a little worried about the six- or seven-year-old kid who must have been flying around the flatbed as the truck jumped over the road's bumps and grooves. Apparently, Guatemala doesn't have the same laws that we do here in the United States—that is, laws that require kids to sit in car seats until they're old enough to get a learner's permit.

Anyway, according to my calculations, if the churches paid twelve bucks per package for two hundred packages and paid an extra five cents per dollar, then the sum total contribution of the churches to the Carmelita community was about $120 per year (0.05 × 12 × 200). I did this calculation three times and then fell into despair. I knew the village was pretty poor, but seriously, how many pencils, pens, and coffins could $120 actually buy? Luckily, at this point I decided to check my understanding with Juan and Jorge. It took a while for me to explain myself, for everyone to be sure that the translation was accurate (translating mathematical relations and equations is no easy matter), and for Juan and Jorge to understand that I was totally off base. But ultimately they assured me that the churches paid five cents extra *per frond*, not five cents extra per dollar otherwise spent. In other words, instead of paying about twelve bucks per package, the churches paid about forty-two dollars for every package, or about a total of $6,000 extra per year (0.05 × 600 × 200). I can't tell you how relieved this made me; with four communities in Petén selling nearly equal amounts of palms to the US churches, this means that the churches are contributing something like $25,000 extra per year to the Guatemalan forests, which, as the economists will tell you, is no chump change (and this amount doesn't even include the money donated to Mexican villages).

My final day in Guatemala was spent in a series of meetings—with people from the Rainforest Alliance, with Jorge Sosa's ACOFOP, and with a government representative who works for CONAP. During the day, I learned an incredible amount about how the government uses law to regulate the concession communities, how the NGOs assist the communities to harvest sustainably, and how eating Mayan nuts (aka, the ramón that I mentioned earlier) can make you stronger, smarter, and more likely to build architectural masterpieces that will survive for millennia. I'll spare you most of the details of what I learned, but two larger points that came out of our discussions are worth mentioning.

First, it became clear to me only after meeting with the representative from CONAP just how remarkable it is that these harvesting communities in the north were able to transition from serving essentially as just a supply source for big exporting companies in Guatemala City to being their own self-sufficient villages. Prior to about 2005, only these large exporters in the capital were really making any money from the palm trade. Not surprisingly, when the government and the NGOs stepped in around then to help the communities develop environmentally friendly harvesting techniques so the communities could export the palm leaves directly to the United States, the big players in Guatemala City put up a lot of resistance. When the communities, acting along with the Rainforest Alliance, finally convinced the big US importer Continental Floral Greens to import directly from them (a process that took at least six months and many trips between Guatemala and the United States), the big companies in the capital ended up losing a lot of money.

In a developing country where corruption is prominent, the fact that the government stood up for small communities and environmental protection in the face of moneyed interests in the nation's capital is pretty amazing. I asked the CONAP representative what could explain the government's ability and willingness to make such efforts. The representative cited, among other things, pressures from civil society, the peace treaty of the late 1990s that put the government in the position to empower local communities, and complaints from Belize that Guatemalan xateros were illegally entering forests in Belize to cut down and sell their palms. Belize's complaints had attracted a lot of international attention to the problem of palm harvesting in Central America.

A second, and related, point is that the Guatemalan government—specifically CONAP and its detailed legal regulation of palm harvesting—has been an incredibly important part of the success story of these villages. Throughout my visits to the communities, I had been skeptical of whether the government was truly playing a beneficial role in the narrative or whether, as I had suspected, the only real movers were the communities themselves, the marketplace (as supplemented by the US churches), and the NGOs. A couple of times, I had asked Juan and Jorge and others whether they thought that the government was a positive or negative force, and each time, although they might have disagreed here and there with the government's approach, they were very supportive of the government's efforts. Still, though, when I was going into my meeting with the CONAP guy, I was definitely expecting someone who was super-official or grumpy or haughty or dopey or something, but that turned out very definitely not to be the case. The guy was incredibly knowledgeable, helpful, down-to-earth, and concerned about the quality of the environment and the people's well-being. Of course, it's always possible that I was hoodwinked by some very good acting, but my strong impression coming out of the meeting was that the government is very much a force for good in Guatemala, at least on this particular issue. Score one for the law!

The westward drive from Flores to San Cristóbal, Mexico, takes about ten hours. Juan drove me the first three hours, and then some guy in a tiny boat ferried us over a river to the border station, where I met Oscar, who was from the Mexican NGO Pronatura Sur and who drove the next seven. Here are some of the things I saw on the drive on the other side of the border:

- More than one dog with only three working legs
- Several very large, open fires
- Lots of military guys with machine guns rifling through some guy's trunk
- Occasional Zapatista graffiti
- Very little kids walking on the side of the crazy, curvy road, even in the dark
- Women and children carrying sacks of wood on their backs with a belt-like contraption strapped to the top of their heads
- Grand, glorious vistas of forests and valleys and hills
- A kid riding a mule
- Many old VW Beetles
- No traffic lights

This was just the stuff on the side of the road. The road itself is also worth a mention. You would think that the unpaved Guatemalan road was bad enough, but the road on the Mexico side, paved though it was, was even worse. Not only is the road curvy and only two lanes, so there's constant passing around slower cars and near crashes into oncoming cars, but wherever there is any semblance of human settlement (that is, almost everywhere), people have built speed bumps in the road to slow down the cars. As a result, driving on this road (and most of the other roads I experienced in Mexico) involved driving fifty miles per hour for about thirty seconds, then decelerating to about two miles per hour to clear the speed bump, and then reaccelerating back to fifty for the next thirty seconds, and so on, over and over, nonstop, *for seven hours.* By the end of the ride, my insides were like slush.

At my hotel in the beautiful old town of San Cristóbal, I met with Romain Taravella, the head of the community forestry program at Pronatura. We spoke in the courtyard, where I was shivering, because the city is seven thousand feet above sea level and I, like an nitwit, had simply thought "Mexico: hot" and failed to bring anything warmer than a T-shirt. Romain is a French guy who speaks a bunch of different languages, has lived all over the world (he did field research for his doctorate in Brazil), and speaks a mile a minute. He explained to me the details of the trip-within-a-trip that I would take the following morning—two days and two nights at two village communities in the Sierra Madre range along with a Pronatura representative named Lubenay and Cesar, a translator from a nearby university. Since the villages are many hours away from San Cristóbal, we would not be returning to the city at night. I asked where I'd be sleeping, and he said he wasn't sure, but probably they'd put me up in somebody's house somewhere in the villages.

We got on the road early the next morning on our way to Santa Moreno, the smaller of the two villages but the one that has been most influential when it comes to palm harvesting in the area. Lubenay didn't speak any English, so we did not communicate very much, but Cesar and I quickly hit it off. A herpetologist who enjoys wading through deep water at night to count toad croaks, Cesar is a motorcycle-driving, ponytail-sporting twenty-something from an academic family. He speaks perfect English and does work for several environmental groups in Chiapas. Cesar and I spent pretty much the entire next two days together, not just talking with various people in the villages but also hanging out, chatting about Mexican and US politics and society, life as an academic, what it's like to be an only child, music, frogs, and lots of other topics. At one point, he found a huge toad with a tick on its foot and picked it up and showed it to me.

But back on our first day, after a drive of about 4½ hours, we arrived at the village, where we were met by the patriarch of the community and, indeed, of the entire palm-growing industry in the Sierra Madres. Luis Corzo Dominguez—or Don Luis, as everyone calls him—is a big man and can accurately be described as a character, a real piece of work. The day we met, he was wearing a straw hat, jeans, boots, blue-tinted sunglasses, and a stained, brown-striped Izod shirt. Loud, proud, and swearing like a sailor on leave in a foreign port, Don Luis showed us around, pointing out a large greenhouse where he plants palm seeds (when the plants reach a certain height, they are transferred into the forest). We saw the bodega, where his granddaughter oversees the women who are sorting the palms brought in by the harvesters in the field, and a huge coffee-bean-sorting machine that has been lent to the village by Starbucks. The area around Don Luis's house is like a palm museum. Right at the entrance, Cesar recognized a type of palm plant that he said hasn't changed in millions of years—apparently an endangered species that is basically a living fossil.

Inside the house, we arranged ourselves around the long dining room table to hear Don Luis tell us the history of the village. Much of what he told us was later corroborated by the current head of the Biosphere Reserve in which the village is located. I should say a couple of things about listening to Don Luis tell the story. For one thing, Don Luis really likes to talk and really does not like being interrupted (by, for example, Cesar, saying something like "Can I please now have a chance to translate the five complete paragraphs that you just spoke at lightning speed?"). So the fact that Cesar was able to translate anything that Don Luis was saying at all was incredibly remarkable and a testament not just to Cesar's ability to understand and speak both languages with skill and nuance, but also to his capacity for remembering long, long tracts of sentences and ideas at a time. Second, the room was incredibly fly-filled. Don Luis's wife gave us each a little glass of something to drink—I think it may have been some sort of iced coffee—but although I drank a little of it, and although I know it was probably rude of me not to finish it, I had to stop drinking because the number of flies that were walking around the lip of the small glass was simply overwhelming. At one point, I counted twelve flies in and on the glass. It was downright *Amityville Horror*-like in there.

The story, in short, is this. Between the late 1970s and the early 2000s, the palm trade in the region was dominated by a guy named Pablo, a so-called coyote who trafficked in illegally harvested plants. Anyone who cut down palms in the forest had to sell them at an incredibly low price to Pablo, who had bribed the authorities to leave him alone and who would then export the palms to the United States and elsewhere at a huge profit. As in Guatemala, Pablo paid for volume, not quality, so the only incentive for the low-paid harvesters was to cut as much as possible as fast as possible. As a result, the harvesters were completely decimating the forests.

Don Luis recognized this problem, and so did the director of the reserve at the time, a guy named Carlos Pisana Soto, who by all accounts was a true hero to the local population. Don Luis and Soto worked together to try to find new markets where the harvesters could sell their palms for a higher price, thus cutting out Pablo. They also worked to figure out how to grow palms from seeds in a greenhouse, something that was surprisingly difficult but ultimately essential to reforesting the area and creating a sustainable yield of palms for the future. Eventually, they solved the seed problem, and somewhere around 2004, Don Luis's son Laison Corzo Montejo (who is now the head of the palm-growing group in Santa Morena) grew the first successful new palm field in the forest and convinced others in the area that they could do the same. Meanwhile, the government and Don Luis convinced Continental Floral Designs to enter into a contract with the village to supply the Texas importer with palms directly, thus circumventing Pablo.

None of this was easy, and in fact both Don Luis and the current director of the reserve (Alexser Vázquez Vázquez) told me that Pablo and his people shot at government officials on at least one occasion. When the contract with Continental was finally signed in 2005, it was a day of great happiness for everyone in the region (well, except for Pablo). As Vázquez told me later in my trip, the day they signed the contract, he wanted to jump for joy. Now, Santa Morena and several other villages in the area have organized themselves into something of a cooperative that meets regularly to share information and promote their products.

I thought this sounded like a great success story, but once again, I was unclear on whether the religious groups in the United States had anything to do with it, so I asked Don Luis about the churches. It was clear that he was grateful for the extra money the churches provide; he mentioned that the village uses the money (which is about the same amount as what the villages in Guatemala receive) for important infrastructure improvements like plumbing and water supplies as well as to help older residents and various other good things. And it was also clear that he has a lot of respect for Dean Current, who he described as a "pilgrim spreading good deeds around the world." But Luis was snarky about the churches themselves. He said that the churches "realized they were covered in gold with fancy cars and houses while the world was burning up and the forests were shrinking" and were looking to provide a little help (I verified on the spot with Cesar that this was an accurate translation). Later, when I asked Luis how the churches should feel about their efforts, he told me that the "gringos should be happy that they have helped the forest." So, there's that.

While we were talking, a thirty-something guy wearing jeans, a pink polo shirt with a surfing logo, sandals, and a Nike baseball hat walked into the dining room. This was Luis's son Laison. When Luis was done talking, Laison led us into the woods to take a look at some palms. We hiked maybe a quarter of a mile, stopping on the way to take in an extraordinary mountain vista with the Pacific Ocean in the distance, before coming to the fields of palms, spread out underneath the forest canopy, different areas of the fields marked with various orange and yellow ribbons to mark their ownership and age. Laison led us through curtains of six-foot palm trees, the smell of lemons and sweat and the buzzing of mosquitoes in the air, until we got to the very first field, the one that Laison successfully planted with seedlings first cultivated in a greenhouse, the one that convinced everyone around that it was possible to grow new palms in the forest. He explained that, twenty years ago, the lush forest we were standing in was basically a burned-out cow pasture; now you could hardly see fifty feet in front of you. The village's policy is to plant the palms as densely as possible, so the planters and harvesters don't have to intrude further up the mountain and so the forest is protected against erosion. Laison was obviously exceptionally proud of what he had accomplished here, and with good reason. As we were about to hike back out, he pointed to a collection of seeds that were stuck to the side of a large tree. "The seeds are so grateful to the mountain," he said, "that they grow off the sides of the trees."

Back at Don Luis's house, after a delicious dinner of fried fish, beans, and plantains, all prepared by Don Luis's wife, we were all sitting around outside talking when a small girl walked by carrying a dead chicken. A few minutes later, she returned, chicken-less, but holding a small, pretty flower, which she presented to me. The girl was Laison's daughter. When she heard that I had studied a little Spanish, she went and got one of her schoolbooks and started teaching me some more. It was incredibly cute and endearing. Soon after (although not until Don Luis had shown us how he manufactures insect-repellent fertilizer with cow manure, worms, coffee grounds, and ants), Cesar and Lubenay and I said our farewells. It turned out that we were going to spend both nights in the other village—Tierra y Libertad, where Lubenay had grown up. Before we left, Don Luis hugged us and gave us some inspiration: "The way I got this contract with the US importer was not to be afraid to try. You should do the same."

Three hours later, we arrived in Tierra y Libertad, a relatively large village of about twenty-four hundred people, many of them youthful and dynamic. If Santa Morena is the old guard of the Sierra Madres, then this place is the young gun. Romain and others believe that with its innovative farming and marketing techniques and the diversification of the products it harvests and sells, Tierra y Libertad is on the cutting edge of forest management in the region. It was late when we got there, and after maybe a half hour of chatting and eating tamales with Lubenay's family, Cesar and I were shown to our lodgings—a concrete house that belonged to Lubenay's great-aunt and was situated at the edge of town. The great-aunt's husband had gone to the United States for a while to pick oranges and watermelons, so she had returned to her parents' house, leaving this one open for visitors like us. The house was partly unfinished—apart from two completed rooms, there were two other rooms that were halfway done and still had dirt floors—but it worked out fine for us.

One weird thing about these two villages was that they were both dry. It's apparently okay to drink alcohol, but there's nowhere to buy it, and the nearest town where it's available is a long way away. This wasn't a problem the first night we were there, because I had had the foresight to bring along a small bottle of scotch that I had bought in Guatemala. Cesar and I shared the rest of the bottle and talked about the perils of childhood bullying and all sorts of other things before going to sleep under the cartoon-character blankets laid on both of our beds. The next night, however, we were out of luck. Even though Lubenay drove us to a little nearby place where it had previously been possible to purchase beer sometimes, the place had recently been shut down by the village authorities. The dryness of the towns is related to their conservative, religious nature, which surprisingly, at least to me, is largely of the Protestant rather than the much more typical-for-Mexico Catholic variety. In Tierra y Libertad, for instance, there are three churches, but only one is Catholic. The others are Pentecostal and Adventist.

The next morning, we were woken early by cock-a-doodling roosters and banging construction workers. After breakfast, Cesar and I hiked up a big hill and then pretty deep into the forest with the village's palm group captain—a youngish, earnest guy named José Arbey Gomez Garcia (he goes by Arbey). He was wearing a soccer shirt with Brazilian colors but bearing the name of the village, jeans, and big white boots, which seemed to be the footwear of choice for palm harvesters. Arbey basically gave us a tour of the forest. At our first stop, he described and demonstrated in detail how trained harvesters approach any given palm plant. One of the leaves of every plant is clearly different from the rest—it is called the heart—and harvesters are not supposed to touch it at all, at least until a separate part of the plant, a smaller and newer stem known as the *vela*, or candle, is large enough. Only one or two leaves of the plant— generally the best-looking ones, with near-perfect *scissors* (the part at the top of each frond where there's one leaf pointing one way and one the other way)—should be taken at any one time. I asked Arbey how many times per year a harvester can take leaves from any given plant. He said that it depends on many factors, including temperature, soil conditions, humidity, and the like, but that generally a harvester can come back to the same plant two to four times per year. There are handbooks and programs to train harvesters in these techniques, and if harvesters do something wrong, like cut a vela or the heart, they can be sanctioned.

The first stop was terrific, but the next stop turned out to be even better. A little more hiking brought us to a 150-square-meter, rectangular field of palms where every single plant was marked with a red tag. Arbey explained that the government requires that every palm field within the forest be accompanied by a small control field where the harvester—with technical assistance provided by Pronatura—collects and records data on every plant (size, age, number of leaves, amount of damage, length of the vela, etc., etc.). This way, everyone knows exactly how well the crop is doing, and the growers can predict how many fronds they can harvest in the coming season, which is an incredibly important thing to know when you are making future sales agreements. The NGOs also use the data to determine how much shade and plant density and other factors are optimal for growing, so they can advise the growers about the best way to increase their yield. Arbey pulled a sheath of paperwork out of his backpack and showed us a page that corresponded to one particular plant we were looking at. The sheet tracked all the relevant information over time in a detailed chart, filled with statistical information. What a great development, I thought, and how far the place has come since the days when harvesters would cut down trees with abandon just to get a couple of pesos from Pablo.

After the hike, we ate lunch and rested, and in the late afternoon, we went to the village's bodega de palma to talk with a group of men who do the actual palm harvesting out in the fields every day. The idea was that I should hear from the workers and not just the people who, like Arbey, are in charge. But Arbey was there at the bodega, changed into a purple striped sweater for the meeting, and he actually did much of the talking. I was never quite sure how freely the other men felt in talking; they were fairly reserved. They did all agree that life as a harvester now is much safer than before, primarily because with the new fields, the harvesters do not have to go nearly as far into the forest, where a person was more likely to encounter a big, angry cat or fall off a cliff. I tried to get the harvesters to talk about whether there was any friendly rivalry among the four palm-growing villages in the region, but no good stories about intervillage football games or whatever were forthcoming. I kept thinking that the meeting would have gone better if we had all been drinking beer.

But nobody was drinking beer, and instead of telling stories of fun and games, the guys were far more interested in making sure I knew how serious their work was and how much more help they could use from people in the United States and elsewhere. Arbey started things off by giving a fairly long speech to that effect and then pointing out that although he is happy to show researchers his work (I was hardly the first person to come down here to learn what the villages are doing), he really does not like it when the researchers do not send what they've written to the villagers themselves. Then a tall guy wearing a blue dress shirt and sporting a mustache who had come in late to the meeting gave an even more impassioned plea for help. I told him I would put his words in the book verbatim (obviously presented in translation and with inevitable errors of transcription):

> We are busting our asses, working really hard because the international community wants us to preserve the rain forest, and we are doing so. We don't use the traditional ways of harvesting. But we're not seeing the profits. The largest cities of the state [like Tuxtla Gutiérrez, the capital of Chiapas] use our water. They have this water because we are defending the mountain. They don't seem to care—they pipeline our streams and rivers, and we don't see profits. We are happy to be protecting our forests, but I think we should have a better life for what we are

doing for the environment. I hope that this kind of project [i.e., this book here] will make people more aware, more willing to pay to support what we are doing.

Now, I'm not particularly good at being "earnest," but it was clear that this man's plea called for a heartfelt response. So I did the best I could to communicate through Cesar my impassioned gratitude to the guys in the room for the work that they were doing, and I offered my commitment to try to spread the word about their efforts (not to mention a promise to send them a copy of the book). My promise wasn't actually that hard, because I meant everything that I had said. I think the men were satisfied, because they said I could be their ambassador for what they were doing. When I got home, I wrote a short piece quoting the mustachioed gentleman for *Slate* and encouraging churches that don't use palms from Eco-Palm to start using them. I can only hope that the article, and now this chapter, will alert at least some people to the great things that these men and women in Mexico and Guatemala are doing for the critical resources that are necessary not just for them, but for all of us. There, how's that for earnest?

The next day, back in my hotel in San Cristóbal, I enjoyed the luxury of warm running water again by taking a long shower and then headed off to Pronatura Sur headquarters for a series of meetings. I met at length with Alexser, the director of the agency in charge of the Mayan Biosphere Reserve in Mexico. Alexser, who was as impressive both in his knowledge of and his commitment to environmental protection and the local communities as his counterpart in Guatemala, explained to me the details of how Santa Morena was able, with substantial government help, to break away from Pablo and become an independent, sustainable palm operation. I also met once again with Romain. He underscored that "the government-private mix is one of the big parts of the story," but he was also quick to point out that the long-term success of the project is not yet assured. Among other things, the Chiapas communities face stiff competition at least in the national market from cutters in other states like Veracruz and Oaxaca, where the cutters do not harvest sustainably.

This possibility of future failure was also the theme of my final interview, which was with Rosa Maria Vidal, the deputy director of the Pronatura Sur office. Vidal, who has worked in conservation for over twenty years, exudes a quiet brilliance that had me sort of entranced. We talked for a while about the relationship between religion and the environment, and she shared another local example that involves the gathering of rare flowers for religious purposes. Mostly, though, she talked about her hopes that religious groups will become more involved than they have been so far in supporting communities like Santa Morena and Tierra y Libertad. "We are looking to see if the religious groups will continue to increase their demand and their premium," she told me. "We're still looking to see if this type of harvesting is sustainable for the long term." According to Vidal, individual churches have been happy to participate during the Palm Sunday season, but the umbrella religious organizations have not demonstrated the kind of year-long, ongoing commitment that may be necessary for the sustainable harvesting practices to continue in the future. Vidal worries that the churches might be more interested in the social benefits of their contributions than the environmental benefits, and that only by fully appreciating the environmental benefits of programs like those in Chiapas will the churches want to increase their commitment. In other words, Vidal was wondering how "eco" the EcoPalm project really is. It's a good question, and one that religious groups in the United States and elsewhere would do well to seriously consider.

One unique aspect of environmental harm is that often the injury is felt far away from the actions that cause the harm. There are several mechanisms that account for this potential distance between activity and injury. For one thing, pollution travels. A pathogen emitted into a river can flow downstream and cause disease hundreds of miles away, or a factory smokestack may be built so tall that the smoke it discharges will affect the air quality in a different state or even a different country.

Second, as ecologists have been explaining for decades, because environmental systems—watersheds, ecosystems, and the like—are often deeply interconnected, harm to one part of the system can cause further harm in a different part of the system. One species goes extinct, for example, and this affects both the species it ate and the species that ate it, and so on.

Third, because of the interconnectedness of the world economy and the growth of international trade, demand for a product in one area of the world that is satisfied by a supply of that product from another part of the world can cause environmental harm far from where the product is consumed or used. Rhinoceros horns are in great demand in Asia for use in Chinese medicine, but the harm to the various species of rhinos is felt primarily in Africa, where most of these animals live.

Finally, and maybe most intriguingly, because many people value a healthy environment even in places where they do not live and will perhaps never even visit, they can suffer harm from environmental damage that occurs far away from them. A good example of this larger worldview that people have can be seen in the debate over whether the US government should allow drilling in the Alaska National Wildlife Refuge. Few people have any plans to visit that remote and inhospitable place, but many nonetheless value it greatly and will fight to ensure that it remains pristine. Economists call the value of environmental goods that do not depend on their use *existence values* and have even developed techniques to try to measure exactly how much people value, say, the existence of a beautiful lake or the continued vitality of a species of crab.

The depletion of palm forests in Guatemala and southern Mexico as a result of Palm Sunday celebrations in the United States and

Europe is a clear example of the international market phenomenon described above—demand for a product in one place causing environmental harm in a different place. And although this kind of mechanism for environmental harm may not be the most common way that religious practice harms the environment, it clearly happens. The gloomy example of elephants being slaughtered in Africa to satisfy the demand of religious believers in China and the Philippines also works this way. But the EcoPalm program also demonstrates that people value a healthy environment even when it's not their own immediate environment. As Dean Current told me in his office in St. Paul: "No one says that they don't want to pay twenty more dollars for sustainability." People in the United States value the health of the palm forests in Central America and will pay to keep them healthy. And herein lies an important lesson for trying to reduce the environmental harm from religious practices: governments and NGOs concerned about the environmental harm caused by religious practices can take advantage of the fact that people—often affluent and resourced people—care about the quality of the environment not just where they live, but everywhere.

In the palm example, of course, the people and institutions donating money to protect the forests are the very people and institutions that are contributing to the damage in the first place, but that certainly does not have to be the case. With the elephants, for example, a number of NGOs raise money to do all sorts of things to help the cause—protecting habitat, tracking and monitoring elephant populations, keeping an eye on court proceedings against poachers, providing aerial surveillance to help wounded animals and lead authorities to law violators. And of course, the money these organizations receive does not come from the ivory dealers or idol collectors who are driving the demand. The same situation holds true for the Ganges River. Several NGOs exist to collect money for projects to help the river through waste management, restoring water flows, planting trees to combat erosion, and promoting education. The government has even recently created a fund for the same type of projects, targeting not only residents of India but also people in the United States, the United Kingdom, Singapore, and elsewhere, who care about the health of the world's most sacred river. The point is that environmentalists can and should capitalize on the worldwide demand for ecological protection when that demand can help counter the negative environmental effects of certain religious practices.

What the palm example also demonstrates quite clearly, however, is that even when foreign aid is helping to counter environmental harm caused by religious practice, the people at the site who are actually designing and implementing the solutions are the ones most responsible for protecting the environment and must therefore be respected, recognized, and rewarded for their efforts. Writing a check is great. Writing a big check is even greater. But we cannot forget that it is the people who use the money to do real work on the ground who are doing by far the most important work—the people in India braving disease in the brutal heat to replant trees or pull waste out of the river, the men and women in Kenya flying helicopters in dangerous conditions to save wounded elephants, the mustachioed man in the blue shirt in Tierra y Libertad who does backbreaking work every day for tiny pay to save the forests for all of us. While most of us sit at our desks, wringing our hands over the fate of the environment and occasionally giving our credit card numbers to NGO fund-raisers, the people I met in Guatemala and Mexico—Juan, Luis, Arbey, Raina, Laison, Burt Reynolds, and all the rest—spend their days on the front lines of the battlefield to save the planet. We should listen to them and learn from them and celebrate their accomplishments and make sure they are paid what they are worth. We may live a great distance from each other, but we are, in the end, all in this together.

③

INDIAN IDOLS

*The Economic Costs of Regulation
and the Problem of Equality*

I was standing on a beach in western Mumbai, on the edge of the Arabian Sea, just me and about a million other people. Now, please note that when I use the phrase "about a million other people," I'm not using it loosey-goosey-like just to make you think, Wow, there sure were a lot of people there. No, I mean there were truly *a million other people* on the beach. What were we doing there? Enjoying a little fresh sea air? Taking a refreshing dip in the crisp, clean ocean? Not even close. We were there to celebrate the final day of Ganesh Chaturthi, the annual ten-day Indian festival that honors Ganesh, aka Ganapati, the big-bellied, elephant-headed, mouse-accompanied god that Hindus revere as the lord of learning and the remover of obstacles. For ten days, people all over India, but particularly in Mumbai's state of Maharashtra, worship Ganesh idols in their homes and in public spaces. Then at some point— usually on the final day of the festival—they immerse the idols and leave them in bodies of water like rivers, lakes, and the sea, symbolizing the god's departure for another year as well as the fleeting and ephemeral nature of life itself.

As you might imagine, this religious practice is not good for the water quality in India. But perhaps you're thinking, Oh, how bad could it be—a thousand little idols left in the water? In fact, in Mumbai alone, over 180,000 idols are immersed every year, and although most of the idols are kept in households and are only about two feet tall, at least 10,000 of them—the ones that are put on public display by temples, neighborhood associations, service organizations, corporations, and the like—are significantly larger. Many stand over six feet tall, and a substantial number of them are twenty, twenty-two, even twenty-five feet high. These things, in other words, are basically elephant monsters. Add to this that the idols are generally painted (often with toxic paints), covered with glitter and other shiny materials, and adorned with garlands and other decorations of all kinds, and you can understand why the ritual of idol immersion—spectacular as it may be—has done a really serious number on the waters in and around Mumbai.

I had arrived about a week earlier and spent my time trying to see as many of the Ganesh idols around the city and learning as much about the festival as I possibly could. I had been looking forward to this final day not just since I'd arrived, but really for as long as I'd been thinking about writing this book. What would it be like, I wondered, sitting in my ergonomic office chair in a tall Boston building, to stand on an impossibly jam-packed Indian beach as groups of people dragged and carried twenty-five-foot elephant-god idols through the crowds and into the sea? Well, I was about to find out. I'd been on the beach most of the day, but during the late morning and early afternoon, the action had mostly involved small families and groups with their modest two-foot-tall idols. Interesting, sure, but not particularly dramatic. As the sun began to set in the early evening, though, the larger idols, having been paraded all day through the densely packed streets of Mumbai, started arriving at their final destination.

And so I stood on the beach, my back to the sea, maybe halfway between where the crowds started and where the ocean began, and looked back up the sandy incline to see four of these gigantic things poised at the slope's edge, getting ready to make their ultimate journey into the ocean. The one on the far right was draped in purple robes and golden necklaces and was holding two unidentifiable objects (was that a CD balancing on his right index finger? probably not) in two of his four hands. On the far left, set back a bit from the others, lurked one that was grasping a tiny bouquet of flowers and sporting what looked to be perhaps a giant ruby necklace around his neck. The one directly in front of me was sitting on some sort of a stool or tree stump, his right knee sticking up into the air in a strange and seemingly uncomfortable position; he was backed by a huge, three-pronged, silver design that made it look like he was leaning against the top of an enormous elaborate dinner fork. The fourth Ganesh, the one directly to this one's right, was the most striking of all. He had eight hands instead of the typical four, each holding a different item (a sword, a scepter, maybe a bow), and behind him towered five cobra heads, looking angry and eager to strike anyone who might impede their idol's progress to the sea.

I stood and watched these things with awe. At the same time, though, I wondered how they were going to make it through the crowd to the water. We were packed together like the mosh pit at a Metallica show. I didn't think the Ganesh idols were going to have a lot of room to maneuver through us. And then the twenty or so people who were in charge of the purple idol started bringing the twenty-foot god down through crowd. Not slowly and carefully, either. These guys had broken out into a sprint. The spectators in the idol's way had to get out of there, and quick. Suddenly a wave of people started pushing and scrambling every which way to avoid the rushing Ganesh, and I found myself being pushed and pulled along with them. I struggled to stay on my feet, as my face was mashed up against some guy's sweaty neck. Along with everyone else, I craned my head the best I could to get a glimpse of the idol as it was being carried past where we were standing, but my view was obstructed at best.

Before long, the crowd stabilized in its new position, people even closer to each other than they were before. And then the idol with the angry cobras started coming down on our left. The crowd, which had just pushed left, was now pushing back to the right. It was chaos. I was sandwiched between three or four other people and carried pretty much wherever they went. At this point, I was genuinely scared. I thought of the dozen or so people who were crushed to death at that Who concert in Cincinnati back in the late 1970s. If I slipped the wrong way, it could happen to me. I wondered what my body would end up looking like after being trampled for another six or seven hours. I grabbed on to the shoulder of the guy in front of me and held on for dear life.

If you're a US citizen who is thinking of visiting India, I have the following tip for you: Do not wait until ten days before your trip to realize that you need to get a visa to visit India. Having waited that long myself (who requires visas these days—aren't we all just one big, happy world family?), I can testify that it can be quite stressful. If someone had asked me the morning of the day before my trip what I was planning for the next afternoon, for example, I would have had to tell them: "Either I'm going on a twenty-hour flight to India, or I'll be sitting on the couch in my living room watching *Two and a Half Men* reruns and eating cheese doodles." The trauma of not knowing if my visa would arrive probably took a year off my life.

Well, it turned out that the visa arrived about eighteen hours before my flight took off, so a mere 7,600-mile and twenty-hour plane trip later, I found myself setting foot in India for the very first time in my life. I've been to about twenty-five countries, some of them extremely different from the United States, and I even lived in mainland China for the better part of a year once, but nothing had prepared me for Mumbai. Formerly known as Bombay (conservative political leaders changed the name in 1996), the western Indian city is one of the most populated and densest in the world, with over twenty million people living in its greater municipal area. It is both outrageously poor and incredibly rich. Delhi is India's political capital, but Mumbai is the nation's commercial center and home to the glamorous Bollywood film industry. But while the city's movers and shakers are busy making multi-million-rupee deals from the back of their Ferraris, maybe 40 percent of the city's population continues to live in slums that generally lack electricity and running water. Only a handful of cities in the world can boast more billionaires than Mumbai, but tens of thousands of children die of malnutrition in the city every year.

Wandering around the city, to get all travel-guidey here for a minute, is an experience that is both exhilarating and exhausting at the same time. The streets are filled with scenes and sounds and smells that you just don't run into elsewhere. Some of the sights, of course, are extremely depressing—practically naked children asking for money, people sleeping in the middle of the sidewalk, human figures horribly misshapen by malnutrition and disease. Others are just surprising or captivating. Women walking around with mangoes, sweets, or buckets of yarn on their heads; the ubiquitous cows, usually accompanied by old ladies who, for a few rupees, will give you a bunch of grass to feed the sacred animals; chai sellers with their big pots of milky tea; street food vendors by the thousands frying up breads, pancakes, and all sorts of other fragrant stuff that fills the air with the smell of a thousand spices; a lady walking a monkey; a kid grinding ingredients into a bubbling cauldron of cumin-scented goo; vegetable markets; flower markets; people wearing clothes of every imaginable type and color; little kids playing marbles; a grain miller grinding up wheat into flour to sell by the bag; groups of young men enjoying games of pickup cricket in the park. It's crazy but unbelievably fascinating. Still, though, between all the people, the cars, the noise, the pollution, and the heat, I found it hard to walk around for more than an hour or two without becoming weary and frustrated and downright nostalgic for the comforts of my hotel room and an ice-cold Kingfisher.

Unlike almost all of the other research trips I've ever taken, I had not been able to arrange meetings or appointments with people before I arrived in India. I had sent e-mails to a fair number of lawyers and activists interested in environmental issues, but the responses I got back, when I did get a response, were cryptic or skeptical and invariably not very inviting. Perhaps I had made some mistake in how I had approached people, but whatever the reason, I ended up in Mumbai without any real contacts. So I was sitting in my hotel room, drinking said Kingfisher, flipping through my guidebook and gazing out my window at the grand view down Marine Drive to Girgaon Chowpatty (the beach where, when last we left it, I was about to get crushed to death), when I figured—maybe I should just pay someone to show me around. Maybe that's what you've got to do in India! So I located a company that puts together custom tours and asked if someone could show me around some Ganesh-related sites in return for a few thousand rupees. Bingo! Within no time, I had arranged such a tour for the next day. I took a nap and then went to a nearby bar to celebrate with a few more Kingfishers and a cigar, all of which was very pleasant until I was forced to sprint home through a torrential rain and lightning storm that left me shaken and thoroughly soaked. Have I mentioned that it was still monsoon season when I arrived in Mumbai?

In the morning, I ate the first of many half-Indian, half-Western breakfast buffets that, between all the lentils, spicy chickpea concoctions, and funny pancakes with lemon pickles and glasses of refreshing watermelon juice, would leave me at least five pounds heavier at the end of my trip than when I'd arrived. Following breakfast, I inadvertently created a small hullabaloo when I walked into the tiny "gym" (a treadmill and a stationary bike) while a Muslim woman in full conservative dress was walking on the treadmill; thirty seconds after I arrived, she stepped off the machine and left. I felt bad.

I met my guide, Vinita, a funny and extremely knowledgeable middle-aged woman dressed in a bright yellow sari, in the lobby of the hotel, and we were quickly off on the first part of our excursion, which was essentially a "best-of" Mumbai car tour. We zipped by the iconic Gateway of India arch and the exquisite Taj hotel, which was one of several Mumbai landmarks that were targeted by

Pakistani terrorists in 2008 (security around the city is still tight in response—I had to go through a metal detector to get into my hotel, for example). We stopped to take a look at the Dhobi Ghat, a massive outdoor laundry where, despite what looked like complete chaos, apparently the city's dirty sheets and clothes are rendered spotless. We also briefly visited the central train station, an incredible Victorian structure, the busiest train station in Asia and the site of the famous musical finale number in the Oscar-winning *Slumdog Millionaire.*

As we sat stalled in the awful Mumbai traffic, Vinita pointed out a eunuch, who was knocking on a car window looking for money, and a couple of Jain businessmen dressed completely in white walking on the side of the road, presumably being careful not to step on any insects since the Jains believe it is a sin to kill any living being, no matter how tiny. When we passed a Zoroastrian temple, Vinita explained how nobody but the Pharisees are allowed into their temples and how it's said that when their believers die, they are placed on a high platform where the vultures can eat their flesh and where the bones can disintegrate and fall to the ground as fertilizer. The highlight of this part of the tour was probably our walk around an ancient part of the city called Banganga, where men were bathing in a central spring that Hindus believe is an actual extension of the holy Ganges River.

The "best-of" tour over, we then got down to some serious Ganesh viewing. Walking around a south-Indian-influenced neighborhood known as Matunga, we stopped in at four or five midsized Ganesh displays that had been erected on the side of the road. Each display typically consists of a Ganesh statue set within a scene known as a mandala. These mandalas can be simple—a few paintings in the background—or unbelievably complex and ornate, like the one I saw the following day; that mandala was basically a two-story palace. Like the idols themselves, the mandalas are all temporary and will be taken down at the end of the festival. Vinita explained that because it is usually cheaper to create the mandala than to buy, decorate, and maintain the idol itself, over time the mandalas have become more spectacular while the Ganesh idols have gotten progressively smaller. Smaller, perhaps, but by no means small. Although the statues we saw in Matunga were by no means the largest in the city, they were still for the most part between six and fifteen feet high.

Because it happened to be one of the festival's minor immersion days, our next stop was a small beach to watch the immersion of some smaller idols. Our driver let us off near a huge, popular green space known as Shivaji Park, where we paused for a few minutes to watch some guys play a game of pickup cricket. Cricket is so ubiquitous in India—among other things, there appear to be at least six cable television channels devoted to the sport—that I found myself constantly wishing I had followed through on my 2008 New Year's resolution "to learn something—anything at all—about cricket." From the park, we walked to the beach, where probably a couple hundred people were gathered around several small Ganesh idols, performing a small ceremony known as a puja, which involves chanting and maybe dancing or clapping and a small fire and various colored powders, as a way of honoring the god. I watched as a couple of brawny guys carried one of the idols into the water, holding the statue above their heads until they were in water deep enough that they could no longer keep the idol out, at which point they dunked it several times and let it go.

Looking around the beach, I saw what ends up happening to the idols. By the rocks on one side of the beach, away from most of the people, the ground was littered with broken pieces of the colorless figures. It was kind of sad. Vinita explained to me that when she was younger, this place had been a real beach where she had enjoyed the sand and water; now, she lamented, the beach has been ruined by the ritual.

Our last stop was the grounds of the mayor's residence, where the city had created an artificial pond that eco-friendly believers could use to immerse their idols in a way that would not harm the environment. As I'll talk more about shortly, many people in Mumbai have become more environmentally conscious both generally and in connection with the Ganesh festival. One step that the government and others have taken is to construct a bunch of these artificial immersion water bodies around the region and to encourage believers to immerse their idols in the pools rather than in lakes, rivers, and the sea. The pool we visited was quite small—about the size of a modest backyard swimming pool in the United States—and would only be suitable for immersing the smallest of idols. A group of maybe a dozen teenage boys in blue and yellow uniforms were manning the pool to help anyone who might come to immerse an idol, but when we were there at least, not a single idol was anywhere to be seen. The contrast with the noisy, chaotic beach scene we had just left was glaring. We talked briefly with the teenagers who reported that they made three hundred rupees a day (about five dollars) for their work, and they claimed that in fact many people had come to immerse their idols earlier in the day. I wasn't so sure. I admired the idea of the pool and the optimism of the kids (they told me that "this way is better"). But I wondered, not for the last time during my travels, whether the highly controlled, largely sterile atmosphere that the environmentalists had set up was really compatible with the vibrant religious practice of the fervent believers who were bringing their idols to the sea.

As extraordinary religious festivals go, Ganesh Chaturthi is actually fairly new on the scene. Nobody knows exactly when people first began worshipping Ganesh during the Hindu month of Bhadra (roughly August–September), but it's clear that the practice dates back for hundreds of years. Until the late nineteenth century, however, this worship took place primarily within the home. It was not until 1893 when Bal Gangadhar Tilak, an activist for Indian independence and a member of the Indian National Congress, took the holiday public as a way of bringing the citizens of Maharashtra together (he himself was from Pune, a city not so far from Mumbai) in opposition to British rule. He began the tradition of installing large outdoor idols of Ganesh for people to worship publicly, as well as the tradition of immersing the idols in public water bodies at the end of the festival. With his support and encouragement, the festival grew substantially in the state over the following decades and even spread beyond into other areas of the country and the world.

Only in recent years, however, has there been any concern about the festival's environmental impact. For one thing, scientists have recently started studying the effects of idol immersion on water quality. A group of researchers in Bhopal, for instance, studied the effect of Ganesh immersions on a large lake used for potable water supplies and found that the practice significantly increased lead and mercury concentrations there. Another study demonstrated that the immersion of Ganesh idols in the Tapi River in the state of Gujarat had led to "major changes" in oil, grease, alkalinity, and other pollutants. The study found that "the plaster of paris, clothes, iron rods, chemical colors, varnish and paints used for making [Ganesh] idols deteriorate water quality" and enigmatically concluded: "No one can stop these religious activities but awareness [of] the people and society can reduce the pollution." A third study found similar effects in a lake in the southern state of Hyderabad. These studies, and perhaps others as well, have led the nation's Central Pollution Control Board to conclude that "in addition to silting, toxic chemicals used in making idols tend to reach [sic] out and pose serious problems of water pollution. Studies carried out to assess deterioration in water quality due to idol immersion reveal deterioration of water quality in respect of conductivity, bio-chemical oxygen demand and heavy metal concentration."

Science is one thing, but activism is something else, and recently, several legal and environmental activists have taken actions to address the environmental problems posed by idol immersion. One such person is Bhagvanji Raiyani, one of India's top public interest lawyers. Raiyani, who is in his seventies, and his organization Janhit Manch, have brought scores of so-called public interest litigations (PILs) and have won many of them. The group's success is quite a remarkable record, given the difficult issues Raiyani has taken up—corruption, education reform, and reform of the judiciary, among others. In 2002, Raiyani turned his attention to water pollution and filed a PIL in Mumbai seeking a ban on the practice of idol immersion. Raiyani's argument was based not only on legal authority, but also on religious authority, which makes the PIL different from anything we might see here in the United States. According to the Bombay Supreme Court, "The petitioner has taken us to various religious scriptures to [show] that the immersion of idols in river cannot be justified on any count. . . . He also contended that Ganpati Bappa, so merciful and kind will hate a [puja] destroying marine life."

In its 2008 opinion, the High Court refused Raiyani's invitation to declare what the Lord Ganesh would or would not hate: "This Court will not be in a position to decide about what a religion permits and what a religion does not permit." On the other hand, the court did recognize that the question of how to manage a conflict between religious practice and environmental protection is a difficult and sensitive one. In a passage that could perhaps be used as an epigraph for this book, the court noted that "the public sentiments and the public interest sometimes do not go hand-in-hand and at times it is found difficult to reconcile between [the] two and it is always necessary in such situations that the Government handle the matter in a way which would ultimately serve the public purpose." In response to Raiyani's petition, the government conceded that "nobody can justify the pollution of atmosphere or of water bodies," but claimed that it was currently developing immersion guidelines that would help alleviate some of the more dangerous effects of the practice. The court was satisfied with the government's representation and disposed of the petition, but it made clear the following: "We expect that the Central Government will consider laying down of guidelines for immersion of idols. . . . Both the Union Government as well as the State Government shall consider it expeditiously because the time lost involving the pollution might prove dangerous for [the] environment of the country in [the] long run."

The government stayed good to its word. Within a couple of years, both the Central Pollution Control Board in Delhi and the Maharashtra Pollution Control Board in Mumbai issued guidelines for celebrating the Ganesh festival in an environmentally friendly fashion. The guidelines encourage the use of nontoxic colors for painting the idols, natural items such as "Nariyal, Supari, Thread, Leaves, Flowers, Milk, Curd, Honey, Ghee, Gangajal" in the puja ritual, and smaller-sized idols. The recommendations also call for a public awareness campaign to educate people about the ecological risks of idol immersion and urge the removal of decorative materials like flowers, jewelry, and clothing prior to immersion.

Finally, the guidelines encourage the use of clay idols over those made with plaster of paris. This last point is particularly important. Ganesh idols were originally made out of clay, an all-natural substance that dissolves in water and causes no negative environmental effects. But clay idols are heavy, difficult to make, fragile, and expensive. As a result, sculptors prefer to sell idols made out of plaster of paris, a much lighter, cheaper, easier-to-use, and more durable substance that happens to be a lot more dangerous for the environment. Many who want the Ganesh festival to be greener have urged the use of clay idols, and these activists have indeed persuaded some sculptors and worshippers to re-embrace clay. On the other hand, many celebrants are resistant to clay idols. Not only is it impracticable to make a large idol out of clay (much less drag such a heavy thing through the streets of Mumbai), but the increased cost would make it more difficult or even impossible for many poor families to buy an idol. This latter concern, as I'll talk about more at the end of the chapter, is significant from an equality and religious-liberty perspective. There is no reason why only people with financial means should be able to participate in a religious ritual.

Crucially, both the national and the Maharashtra guidelines take the form of suggestions only. They do not impose any requirements or prohibit anyone from doing anything. Given the popularity of the immersion ritual and the sensitive religious character of Ganesh worship, it is not surprising that the government did not try to regulate the practice through actual legal rules. A few areas around the country, however, did try to go further. The state of Gujarat may be the most prominent example. There, the state's Forest and Environment Department prohibited the use of plaster of paris for making idols: "Idols should be made from natural materials as described in the holy scripts." The state also prohibited the "use of toxic and non-biodegradable chemical dyes for painting idols." The experiment did not last long. A large group of idol makers challenged the ban before the National Green Tribunal, which found in their favor. Among other things, the court found that only the central government, and not a state government, could issue legal prohibitions or restrictions on the use of certain materials in idol making.

The day after my tour with Vinita, I headed out on my own to the Khetwadi District of the city, maybe a forty-five-minute walk from where I was staying and one of the undisputed Ganesh hot spots around town. I had read that the district contained twelve small parallel lanes, each of which had its own unique idol. Some of the Khetwadi idols—particularly the ones in the eleventh and twelfth lanes—were among the largest, most extravagant, and most famous in the entire country, attracting tens of thousands of visitors every day. I probably should have just taken a taxi from my hotel, but when I looked at the map, it seemed easy enough to get there by foot, so that's how I set out. Ha! Once I left the main drag, the bewildering chaos overwhelmed me, rendering me hopelessly lost and somewhat concerned that I was having a stroke. After a while, I found a little space to take a breather on a corner next to a cow, when a man noticed my distress and pointed me in the general direction of where I wanted to go, and a little sweaty wandering later, I finally found my destination.

I spent the next couple of hours going from lane to lane, viewing with awe this incredible set of magnificent Ganesh idols. Even though the statues all depict the same god, each one is wonderfully different. The idol at the ninth lane, for instance, was advertised as being eco-friendly and was relatively small (maybe five feet tall), made of clay, and posed in a bucolic, natural setting. The seventh-lane idol was displayed inside a simple, dark room glowing with blue light; the Ganesh itself was sitting on a giant scepter, holding a small spear, and looking upward. At the eighth lane, which inexplicably sported some sort of weird Disney theme, a long entry corridor leading to the Ganesh itself was lined with little creepy figurines—a wide-eyed Minnie Mouse out of a bad acid trip; a tiny, screaming, blue cat. The Ganesh itself—pinkish, modestly sized, holding a fan in one of its left hands—was not Disney themed at all, unless its giant ears were supposed to be reminiscent of Dumbo.

The atmosphere in and around these Ganesh displays was more like a festive celebration than any kind of solemn worship—more like a New Year's Day parade than an Easter service. The streets around the parallel lanes were filled with food hawkers and an occasional person covered head to toe with colored powder, but for the most part, life was going on as usual—women were carrying buckets of combs on their heads, men were moving around long metal pipes, a teenager sported a shirt reading YOUR FACE, MY ASS, WHAT'S THE DIFFERENCE? Visitors take off their footwear before going in to see the Ganesh idols, resulting in really dirty feet and a lot of worrying about whether one's $145 Birkenstocks, which one should have left in the hotel and replaced with a $2 pair of flip-flops, were still going to be there upon exiting (they always were). Some of the more popular idols had long lines of people waiting to get in, although for a small price, and a little bit of shame, it was possible to circumvent the line. For example, at the eighth-lane idol—the Disney-themed one that was displayed behind a facade of an elaborate castle—I noticed that some people were paying twenty rupees (maybe thirty cents) to bypass the long line, so even though I felt like a jerk, I paid the money and went right in.

The eleventh-lane Ganesh is one of the most famous in the entire city. A few years earlier, it had hit forty feet tall, the largest in the history of the festival. When I was there, it was only twenty-eight feet high (the government had urged idol displayers to keep the size of the idols down somewhat, and it seemed that the advice had stuck a bit), but it was still allegedly the tallest idol at the festival. The line for the eleventh-lane idol was really long, much longer than at the eighth-lane idol, but still reeling from the guilt of bypassing that line, I decided to save my thirty cents and wait with the masses. That was super fun! I waited for over an hour, crammed in with a sea of people whose sense of personal space is, let's say, not quite as expansive as my own, watching those who paid their thirty cents breeze freely by us, before finally making it in to see the Ganesh.

But whoa, was this some Ganesh!? Displayed under an elaborate, multicolored, oval dome, the gargantuan elephant god was standing on something that looked like a pogo stick, its right knee raised up almost to touch the lower of its four right arms, the left knee bowed out to its left, the trunk hanging down below the lowest of the idol's four left arms, the one holding a tiny shield. In front of the Ganesh, and below it, crouched the fiercest mouse in Mumbai. Every Ganesh has its accompanying mouse sculpture, usually a little friendly looking guy, but this rodent was immense and dressed to kill—wearing only a loincloth and grasping a fearsome spear in its right hand, its left hand balled into a fist, this guy was seemingly ready to punch in the throat anyone who might get past its sharp, bared teeth. I don't know why anyone would want to try to harm an eight-armed, bejeweled elephant god, but it looked like this twenty-eight-foot one wasn't taking any chances.

The only Ganesh that I saw during my stay in India that might have eclipsed the grandiosity of the eleventh-lane idol was the one at the twelfth lane. This one too is famous throughout India, and indeed the year I was there, it won, for the fifth time, the award for best idol in the city from the *Times of India*. Although the idol itself was fantastic—it was sitting, its lower body draped in orange, four arms rather than eight, but covered with beautiful jewels and multicolored fabrics of all kinds—the real attraction here was the setting. The sponsors had constructed an elaborate and enormous two-story palace that must have been forty feet tall. Royal blue walls were graced with golden arches through which a soft purple light glowed. Majestic chandeliers hung from the gilded golden ceiling. Worshippers enter into a narrow mezzanine at about eyeball level of the Ganesh, who sits in a giant throne in the middle of the palace. The mezzanine ends at a staircase that takes you down to the ground level, where the crowd gathers to pay its respects to the idol and also take the idol's picture. There is a mouse, of course, but unlike the eleventh-lane squeaker, this one was silvery and happy looking and holding the mandala's red powder in its little paws. I smiled at the mouse, collected my Birkenstocks, and took one of Mumbai's famous door-less trains—fewer doors means more people can scrunch in, which is good, even if once in a while somebody falls off the train and dies—back to my hotel and yet one more (okay, three more) ice-cold Kingfisher(s).

Here are some of the festival-related stories that were making news during the week I was in Mumbai:

- A famous Bollywood movie star was offered twenty million rupees (over forty thousand dollars) to make a thirty-minute appearance at a Ganesh idol in the state of Gujarat. Bollywood actors often receive money to make appearances with the idols, but this was an especially large amount of dough.
- Two members of a Ganesh mandala in Mumbai employing loudspeakers at 12:15 in the morning assaulted a police officer who was attempting to enforce a noise ordinance. According to the *Times of India*, "this incident shows how lawless a section of mandalas have become. Years of cops looking the other way and refusing to clamp down on noise-norm violations, combined with the arrogance bred by political patronage, are responsible for this situation."
- Some person or organization concerned with city traffic erected a large sign on a prominent part of Marine Drive

declaring: MUMBAI DESERVES BETTER ROADS. LORD
GANESHA ARE YOU LISTENING?

- A video showing several men who tended to the most
  famous idol in all of Mumbai harassing women went viral
  on social media and sparked a public response from the
  state's home minister.
- On one of the early immersion days of the festival, at least
  thirty worshippers were bitten by stingrays at the main
  Mumbai beach. A dozen were taken to the hospital for
  treatment. Signs immediately went up warning immersing celebrants to be careful.
- With the forecast showing rain for the big immersion day,
  a lot of talk centered on how to protect the idols as they
  made their way to the sea. Large plastic sheets were to be
  the preferred method of protection. As it happened, it did
  not rain on the immersion day at all.

The most famous idol in all of Mumbai, the one where the workers
were caught on camera harassing female visitors, is the Lalbaugcha
Raja, or Lalbaug, which is so renowned for its purported power to
grant wishes that well over a million worshippers visit it every day
of the festival. I had heard that it would take several hours of waiting in a serpentine line to get in to see the idol, and I had no interest
in doing that. I did, however, want to check out the neighborhood,
which contained many other idols, some famous in their own right,
as well as just a lot of festival-related excitement and activities. Having had success with my first tour, I arranged another one with a
different company to visit the Lalbaug neighborhood.

My guide arrived at the hotel late and wet, drenched by the
rain and delayed by the trains. Maybe five feet four inches tall,
with long, slender fingers and a bushy gray beard, he appeared de-
cidedly guru-looking. He introduced himself as Ramanand, and
as we jumped into the smelly car provided by the tour agency, he

explained that he was a former union laborer turned sustainable farmer turned guide-follower of a true Indian spiritual leader-Waldorf schoolteacher. By the time we made it to see our first Ganesh, he had already launched into a complicated lecture on the relationship between spirituality, ecology, and sustainability—a speech of which I followed maybe 16 percent.

I did, however, understand his view on the Ganesh festival (and other Hindu immersion festivals), which he finds, in its current incarnation, hollow and environmentally dangerous. As he explained, in the beginning, one can imagine parents teaching their children about the impermanence of life by making a little clay figure of a god, covering it with flowers, worshipping it for a while, and then putting it into a nearby pond where it could dissolve right back into the earth where it came from. But now that people have moved to the cities, they use resources from elsewhere to build and decorate their idols and then put the idols into water far from away from the material's real source, thus linearizing the process, breaking the cycle of life, and rendering the whole practice unsustainable. The festival worries him greatly. He called it a "blind ritual" that many people follow out of fear that if they stop doing what their family has done for years, they will suffer. "We're god fearing," he said, sadly. "Not life loving." Later, when I asked him over lunch whether I would see anything spiritual the next day during the big immersion festivities, he pretended to drink from a bottle, *glug glug*, and said, "Oh, there will be spirits all right—but that'll be the only spiritual thing about it."

After a quick trip to a big Ganesh in the main tourist area of the city, where Ramanand took my picture patting the toe of the idol, which I learned is something that one is supposed to do when visiting an idol, we made our way to the ultrachaotic part of town surrounding the famous Lalbaug Ganesh. A quick check with the police managing the line at the idol revealed that the wait to see it would be at least two-and-a-half hours, so we set out to see some other things instead. We visited a sculptor who was busy working on a bunch of idols of the goddess Durga for the next Hindu festival.

At one point I mentioned to Ramanand that I was hoping to pick up a little Ganesh idol of my own to take home with me, meaning that I would try to find one in a store somewhere. The next thing I knew, he had arranged with the sculptor to bring me up to the top floor of an old building where I was suddenly surrounded by two-foot-tall Ganesh idols and urged to pick out a nice one. I settled on one who was holding an orange and sitting on a big leaf, because it seemed kind of environmentally themed and, with the help of Ramanand, I bartered with the sculptor until he agreed to sell it to me for thirteen hundred rupees, or about twenty US dollars.

(As it turned out, getting the idol back to Boston was an ordeal. Because the idol was heavy, awkward, and easily chipped, I wanted the sellers to pack it up for me in a crate that I could readily take home. The sculptor was fine with the request, but claimed that the "carpenters are celebrating a carpentry holiday and are worshipping their tools and cannot pack it up today." This seemed odd, but with little alternative, I agreed to have them deliver the crate to me the next day at my hotel. When the idol arrived at the hotel, it was in a cardboard box filled with newspaper and coconut fibers, hardly protective of my investment and definitely not put together with any tools that had required worshipping. To bring it home, I took it out of the stupid box and put it in a paper bag. I just carried it like a baby or a pet everywhere I went on the trip— through the security line, into the bathroom, in the backseat of the taxicab—and stored it in the overhead compartments of both my flights. I got a lot of weird looks, but now it's here at home, sitting on the shelf behind me with its orange and watching me type this very sentence.)

Ramanand and I went on to check out a couple of the big idols in the area, including a famous giant one known as Galli. By this time, however, I had gotten pretty much Ganeshed out. So, after those two were done, I called an end to Ganesh viewing, and we headed for lunch, which consisted of delicious fish slathered in coconut gravy and some funny patties made out of the little flowers that you can find inside a banana plant. During lunch and the long car rides to and from the restaurant, Ramanand continued to spin out his philosophy of life. As long as he remained on these abstract topics, I have to admit that I zoned out a bit (I had written in my notes at this point the sentence "Saris are unforgiving, re: side fat," so I guess I was looking out the window). But when he turned back to the Ganesh festival, I perked up.

He said a few really interesting things. For instance, he suggested that when you look at the Ganesh idols, you're supposed to realize and understand the idea that life is change, that all is impermanent. But instead, people in Mumbai just go see the idols to check them off their list—one, two, three, four, five, see them all, have lunch, go home. He explained his view that any person or organization that makes something, like a Ganesh idol, should anticipate its destruction and responsibly provide for it, but the idol makers don't do that, and so the festival, ironically, turns out not to be about what it's supposed to be about. Finally, he said that one of the many names or meanings of Ganesh is "the remover of obstacles." But, Ramanand explained, "look how many obstacles this festival has put up to getting around the city." When I asked him finally whether he was optimistic that the festival might become more eco-friendly, he seemed to think it was unlikely: "It's really difficult. It's a big business, and that will typically prevail. There are a few people out there doing good eco-friendly work, but it is very hard in the face of all the money."

The following day was the one I had been waiting for—the last day of the festival, when all the idols that had not yet been immersed would be sent to their final resting places at the bottom of Mumbai's lakes, rivers, and seas. I had expected that things would start getting going early in the day, but in fact, the beach was basically empty in the morning (I could tell, using my binoculars from the hotel room). So I watched a bunch of live television coverage of the festival instead, as those in charge of the biggest idols started parading them around the streets through enormous throngs of celebrating Hindus.

I headed to the beach about noontime, and for a few hours, most of the action was low-key. The idols were fairly small ones, probably family-owned, and they were accompanied by coteries of worshippers large and small, some gentle, others exuberant, many of the groups all wearing the same color of clothing (and covered with the same colored powder), like teams. The group would set the idol down in the sand and do a little puja ritual, complete with some combination of fire and coconuts and bells and drumming and dancing, and then two men would bring the idol into the water and give it a couple of dunks before handing it over to some red-shirted official workers who loaded all the idols onto an orange raft. When the raft was filled with idols, maybe fifty or so, the workers would take it out—way, way out—into the sea, where presumably they threw the little guys into the water one by one (they took the rafts so far out, it was impossible to see what exactly the guys did with the idols, but the rafts always came back empty).

The scene on the beach was fascinating. Vendors made their way through the increasingly dense crowd to sell all sorts of things— cotton candy, water, balloons, chai, gross-looking ice cream bars from a big metal can. Lots of women and little kids were begging but not too persistently, though one beggar wouldn't stop tapping my foot. Other kids were digging in the sand and making pyramid-shaped sand temples. A lot of people were out there trying to keep the place relatively clean—teams of official workers were dragging around round sled-like contraptions, collecting trash and organic materials that had been taken off the idols before immersion and then separating the trash from the organic stuff and placing them in different receptacles.

After a late lunch at a nearby famous vegetarian restaurant that claimed to have the "Veg Edge," I headed back to the beach to await the arrival of the giant idols. For a couple of hours, I took up a great position up against a fence right at the place where the idols were arriving and just stood there and watched as bigger and bigger ones starting making their entrance. I struck up a conversation with a hip Mumbai guy named Neel who was sporting funky half-red glasses, and we talked about the meaning and origin of the festival. I asked him what the phrase was that everyone had been chanting non-stop all day long, and he told me the phrase was "Ganpati Bappa Morya," which is a cry of praise for Ganesh. Neel and his friend had been walking around Mumbai all day taking pictures of various processions, including the Lalbaug's, so he showed me a bunch of the incredible pictures, but the best one was the picture of his own family idol—a small one, made out of clay, with a maternal theme to honor his mother. Of all the people I talked to, Neel had the most moving perspective on the festival. As he eloquently put it: "There's a competition for the biggest, best idol, but really, whether it's big or small, fancy or simple, expensive or cheap, it's all the same god."

After a while, Neel and his friend moved on, and I made my way back into the middle of the crowd on the beach, which is where I was at the beginning of the chapter—in a sea of people, my back to the actual sea, looking up at the approaching Ganesh giants as they made their dramatic descent toward the ocean. As you may have surmised by now, I was not trampled to death by the crowd. I wasn't even injured. But I did get pretty worried, and so after watching four or five of the idols make their way through the crowd up close, which I have to say was a breathtaking sight, I fought my way out of the densest area of the beach and out of harm's way. Anticlimactic, I know. But what did you expect this book was going to be? Some kind of tragedy, like *Into the Wild*?

For a while I stood in a relatively uncrowded (read: still pretty damned crowded) place near some concession stands and watched the action from there. As the sun set over the bay, one enormous Ganesh idol after another was marched into the sea by teams of worshippers. It was as if the Rose Bowl Parade dead-ended at the Pacific Ocean but then kept on going. I tried to forget about how these things were going to end up lying on the ocean floor creating havoc in the marine ecosystem and attempted to focus instead on the surreal beauty of the scene. With the extravagant, multicolored statues marching in line through the purple sky, disappearing into the dark embrace of the sea, it wasn't very difficult.

The next morning, I returned to the beach to assess the damage. The scene was grim, like a battleground littered with elephant god corpses, a veritable graveyard of Ganesh idols. Slabs and chunks of the larger idols were strewn about, their coconut husk threads giving them away if they happened to be stuck upside down in the sand. Lots of more or less intact little idols were washing up as well, some of them missing an ear or part of a trunk or something else. The paint on the idols' remains had washed off, dissolved into the sea, leaving them as pale gray ghosts to haunt the shore. All around, kids were picking up the smaller idols and putting them in piles or standing them up next to each other, maybe in a circle, like a little sad seminar. I watched as one idol washed up out of the sea and landed pretty much upright, leaning against a pole with a red flag on it. The idol rested there for a minute or two before a big wave came in and hit it, causing the object to flop over unceremoniously in the mud.

I took out my camera to take a picture of a pile of the washed-up idols, but a police officer quickly put a stop to that. He came running over, giving me the signal for *no way*. He was friendly enough, but he was definitely not going to cave. "A picture could be used for the wrong reasons," he told me, as I put away my camera. I considered surreptitiously taking a picture anyway after he left, but much as I would have loved to take an intensive course in the local language, I had no interest spending any time in a Mumbai jail, so I didn't take the risk.

I wandered down to the far end of the beach, where I ran into a group of college students who were dressed in identical blue shirts and picking up trash. I struck up a conversation with one of the guys, who explained that students from the college volunteer every year to do what they can to clean the place up on the day after the immersion. When I asked him if Ganesh idols wash up onto the shore a lot, he said, "Oh yes—they are made of plaster of paris so they don't dissolve in the water." I told him about the cop who wouldn't let me take the picture, and the kid was hardly surprised. "No, no, they won't let you," he told me. "It could become a political controversy."

After ninety minutes or so, I left and wandered back to the hotel. The walk home was a lot quieter than it had been the evening before, when processions of Ganesh worshippers had filled the night with their ecstatic chanting and drumming and dancing and blasting of fireworks from the side of the road. This morning, there was none of that, but I did stop halfway back to watch three little kids, maybe five or six years old, tear apart a small Ganesh idol that perhaps they had made for a school project. They threw its parts—a leg, the trunk, an itty-bitty arm—into the sea, where the pieces disappeared with a series of tiny *plop*s.

With the festival over, the next day I took a trip to Pen, a small town about three hours from Mumbai that is famous for its Ganesh idols. From what I had read about the place, it was filled with artists and sculptors renowned for their work, and idols made in Pen were said to be the best ones to have. I had made an appointment to meet with one of these famous sculptors, Shrikant Deodhar, a fourth-generation Ganesh sculptor known around the world for his idol making and, increasingly, for other types of sculptures as well. He has long been the head of the sculptors' association in Pen, and according to his website, before he retired from the Ganesh-making business, he used to make between eight and ten thousand Ganesh idols every year.

To get to Pen, I hired a driver through my hotel, but unfortunately, nobody had quite conveyed to the driver that I actually wanted to leave Mumbai. Expecting a short job, he became grumpy upon learning that he had to spend the day driving back and forth to

Pen, but a job is a job, and so he reluctantly agreed to make the journey. Once we got to Pen, however, finding Deodhar's house wasn't
so easy. I had a cell phone number for the master sculptor, but when
the driver called it and nobody answered, he got out of the car to ask
directions. The effort to find Deodhar was comical. I watched from
inside the car as a cow and then a woman with a can of gas on her
head passed in front of me. The driver asked a guy with a rickshaw,
but when that didn't yield results, he asked a guy selling apples. But
the apple guy didn't know, so he asked the nearby banana guy, but
he didn't know either, so the driver tried asking a guy who was selling scarves and coconuts. No luck. When the driver got back in the
car, saying, "Confusion, confusion," over and over again, I asked him
if he wanted to try calling Deodhar again. As it turned out, the driver
had mistaken my 9 for a 4 in the phone number I'd given him, and
once that was cleared up, calling Deodhar was no problem, and we
found the house within minutes.

I sat with the sculptor in his white and airy living room, sipping tea and talking about the holiday. A little intimidated at first
(the guy had just come back from Spain, where he had been giving
weeks of workshops on sculpture making), I was quickly put at ease
by Deodhar's good humor and willingness to share his knowledge
with me. He talked about his family and their long history of idol
making in Pen, as well as the history of the festival. I was interested
in learning what made Pen so famous and what made the town's
sculpture so desirable, but at first I had difficulty figuring it out.
Deodhar explained that making a Ganesh sculpture is difficult. You
start with a human body and an elephant face, which is itself kind of
"absurd," but then "you can go your own way—fat, thin, big ears, really small ears, it still looks like an elephant." Like most of his Brah-
man brethren, Deodhar prefers a subtle and sophisticated Ganesh
figure, one with soft colors and delicate features, one whose head
looks like an actual elephant. His clients have not always shared his
refined taste, though, and while it had frustrated him as an artist to
produce more garish Ganesh idols, as a businessman he provided
what his customers wanted, even if it involved putting the god on

a tiger or a very big fish. "The fishermen want a very spicy idol," he explained, "because they are spicy people."

After a while, I steered the conversation toward the environment and the differences between clay and plaster of paris idols. Although Deodhar claimed to care a great deal about the environment, it was clear and not surprising that as the president of an association representing fifty thousand sculptors making plaster idols, his true allegiance runs to his fellow idol makers. He has been in the middle of the controversy between environmentalists and plaster sculptors and, after reading a report issued by the Maharashtra Pollution Control Board, believes that although plaster can block waterways, it is not itself harmful. At times during our conversation, Deodhar's impatience with environmentalists came to the surface. For instance, at one point, he said it was a big joke among the sculptors how minor the pollution from a one-day festival must be compared with the manufacturing and chemical companies that pollute the water all year long. Another time, he told me that out of a hundred environmentalists, ninety are really marketing some environmentally friendly product or service. Only the remaining 10 percent, he said, are true environmentalists, although even among these, many are really just showing off. Ultimately, Deodhar said that he is happy when someone wants to buy and immerse a clay idol, "but if you say as a high society that you can't sell the other type, that's when there is going to be a fight."

It is hard to understand the idol-making process just by listening to someone talk about it, and I honestly didn't really get it until Deodhar ushered me into the studio where he does his work. These days, he is no longer mass-producing Ganesh idols (the administration aspect of running the business was killing his inner artist), but is instead focusing on expanding into other types of sculpture and doing a few, as he calls them, signature Ganesh idols. His studio was lined with shelves and shelves of idols, molds, and other materials, and piles of plaster powder dotted the studio's floor. I asked him about the powder, and he explained that in India, instead of buying premade plaster, you have to buy powder and mix it up with water yourself. As Deodhar explained the many, many steps that go into making a finished idol—at least twenty-five people can be involved in making a single sculpture—I started to understand what he'd meant when he'd said that idol making is some seriously difficult work.

The first step is to create the original model out of clay. This is where the skill of the model maker can make all the difference between a beautiful work of art and, say, a piece of crap, and it's where Deodhar himself excels. He showed me one of his models, and the vast difference between what he had made and the typical Ganesh idol like the one that I bought a few days before was immediately apparent. Most idols look like cartoon characters; his looked like a genuine elephant (albeit one with a human body and a nearby friendly mouse). Once the model is done, someone else will make a mold out of it by applying many layers of latex over the model and then skillfully removing the latex, leaving a hollow mold that can then be filled with plaster over and over to make hundreds of sculptures. Separate molds are made for various delicate parts of the idol, like the Ganesh's hands and tongue. These separate parts are all put together to create the complete Ganesh. This, in turn, enters the finishing phase, which consists of polishing, priming, painting, adding ornaments, and, most interestingly, eye painting. According to Deodhar, the painting of the idol's eyes requires great skill, and there are people whose job it is exclusively to paint eyes. These guys work freelance and can make good money painting a hundred sets of eyes or more every day.

We moved from the studio into Deodhar's spacious office, which was adorned with posters from Ganesh exhibitions, some in foreign languages. He tried to turn on the lights, but when they didn't work, he shrugged it off by saying, "This is India." It was interesting hearing the sculptor talk about how his customers had reacted to his decision to get out of the mass-idol-making business. "People are so emotional about their Ganesh," he said. "They didn't want me to stop, because they'd been getting their Ganesh from my family for fifty years." According to Deodhar, it took him five full years to transition out of the business, because people kept demanding that he provide them with his idols. Even now, although Deodhar has made sure that his customers can get idols from other master artists, he says his former clients sometimes still harass his wife, trying to get her to convince him to make them an idol. "Shrikant Deodhar," they'll say, "your Ganesh used to talk with us. Your friend's idol does not." I can understand their complaints. I've had my own garish Ganesh sitting behind me for months now, and although I like having him there, he hasn't spoken to me a single time.

My week in India gave me a lot to think about. For one thing, it was clear from the situation in Mumbai that the government could not possibly institute a flat ban on idol immersion in natural waters without causing something resembling a revolution. My guess is that the most environmentalists can hope for, at least presently, is to convince the government to impose relatively minor regulations to reduce the most harmful impacts of the practice. Indeed, having never witnessed a religious celebration quite as intense and immense in scale as the Ganesh festival—particularly the final day of immersion—I was struck by just how long a road the environmentalists have in front of them. For those in India and elsewhere who are hoping for significant change, they are probably wise to keep in mind that it's a marathon and not a sprint. Or, in other words, in some cases of extreme religious devotion and fervor, environmentalists in the government and elsewhere should recognize that change could take a long, long time and that it may be best to seek a series of small changes rather than going for too much too quickly, which could risk a backlash and ultimate failure.

The morning-after scene at the beach, where the police officer stopped me from taking a picture of the pile of washed-up, broken idols, made me think about the importance of the free flow of information. If the environment is going to get a fair shake when it suffers harm from religious practice, people are going to have to know about the existence and degree of the harm. Lack of sufficient information to make rational choices is a classic justification for regulation in a market economy—regulations requiring, for example, the disclosure of toxic releases, nutritional information about food products, and possible side effects of medications are justified on these grounds—and it justifies intervention here too. The government has a responsibility to collect and distribute information about the harms to the environment brought about by religion so that the people affected can make informed choices about how to proceed.

In the case of water pollution caused by Ganesh idol immersion, the government in India has taken steps in this direction, as evidenced by the judicially mandated reports and guidelines issued by the Central Pollution Control Board in Delhi and the Maharashtra Pollution Control Board in Mumbai. On the other hand, the government's reaction (in the form of the cop) to me and my camera on the beach was deeply disconcerting. There, the government was not simply failing to provide information itself—it was actually stopping me (and presumably others) from collecting and disseminating information. How can the population of Mumbai decide what, if anything, should be done about the environmental damage caused by the Ganesh immersion if they are shielded from information that is extremely relevant to making that decision?

The issue that I found myself wondering about the most, however, was whether increased regulation of the Ganesh festival would have troubling, unequal effects on rich and poor worshippers. The vast difference between the wealthy and the impoverished is so vividly on display in Mumbai that it was hard not to think about the question of equality in nearly every context, including the one I had come to research. Regulation of any sort that increases the costs of an activity, even if the costs are measured in terms of difficulty of engaging in the activity rather than purely economics, is likely to impose unequal burdens on regulated parties or the people who rely on those parties. If, for example, a state imposes costly regulations on abortion clinics—say, requiring the clinics to comply with the same building and staffing requirements that surgical centers have—then some clinics are likely to close, and if some clinics close, women who seek an abortion will, in general, have to travel longer distances, spend more time, and pay more money to access a clinic. This change may have little effect on women who have plenty of time and money to go to a clinic farther away, but those with fewer resources may actually lose the ability to visit a clinic altogether.

Typically, in the environmental law context, we don't consider the unequal effects of regulation to be a terrible problem. A regulation requiring all zinc processing plants to implement a certain type of costly antipollution technology will likely be more burdensome on smaller, less successful plants than it is on bigger, richer ones. The regulation might even put some struggling plants completely out of business, but this potential downside is often understood to be the inevitable result of regulation and, while perhaps regrettable, is usually justified by the environmental benefits of the new technology. But people practicing their religion are not like businesses making money—they are individuals expressing their most important beliefs about the world around them and are relating to their communities in profound ways. And so the government should be very wary of implementing any regulation meant to protect the environment that will allow some believers to practice their religion but not others.

This lesson is particularly relevant to the question of whether the Indian government should require that idols be made out of clay rather than plaster of paris. In some ways, this is the regulation that would best balance freedom of religious practice with environmental protection. Hindus could still worship Ganesh idols all week long, and they could even carry them through the streets and celebrate wildly as the faithful immerse the idols in the sea. But the immersion would have far fewer negative environmental effects. Clay dissolves, and as long as the idols are not painted with toxic paints, the impact on the water quality would be minimal. Although such a regulation would make it harder, if not impossible, to make and carry extremely large idols, that cost to religion (a real cost, I concede) may well be worth the environmental benefits. Likewise, although the regulation would have a serious negative effect on the businesses of many sculptors—a real problem in a place like India, where many people remain so poor—I'm still not convinced that the costs would outweigh the benefits.

The problem I worry about the most is whether requiring clay idols instead of plaster ones would make it harder or even impossible for many poor individuals and families to have and immerse a Ganesh idol of their own. Since clay is much more expensive than plaster, this requirement would raise the cost of buying an idol to such an extent that they would no longer be affordable for some families, and then the government would have created a serious problem for religious freedom. Some people—those with the financial resources—could buy a clay idol, keep it in their home, worship it, bring it to the sea, and immerse it, while others would not be able to purchase an idol at all. Of course, the government makes all sorts of decisions that have an effect on the general state of economic inequality in a country, and this inevitable inequality will make it harder for some to worship than others. But the government should not make such conditions worse by passing specific regulations that make it harder for the poor to practice their religion. When it comes to religious freedom, we should all be equal.

EAGLES

A Reprise

It's hard to explain what a dead bald eagle smells like. It's not as bad as you might think, but it's not great either. Maybe kind of like a dead fish. Not a slab of sashimi-grade tuna, of course, but also not like something that's been decaying on the beach for a week. Probably a dead eagle smells a lot like a dead chicken. But then again, I don't really know what a dead chicken smells like. Probably it smells something like a dead fish. One thing is for sure, though: processing a dead bald eagle sends a lot of tiny fluffy feathers up into the air. Back in Commerce City, it was practically snowing dead-eagle feathers in the room while I watched two people from Bernadette Atencio's staff work through a few newly arrived birds. It was a little disconcerting that the two men working with the birds were wearing face masks and full-length protective suits while I stood there trying to breathe as little as possible so as not to contract any eagle-carcass-fluffy-feather-borne diseases, whatever those might be, if indeed such diseases exist (and how could they not?).

The first box of eagles was from South Carolina. Immediately, Atencio told me that these would almost certainly be bald eagles and they would probably be on the small side. The word she actually used was "dinky." A young man named Adam, who had been working at the repository for a while before heading off to college, unpacked the box and took out what seemed to me to be a not entirely dinky bald eagle. He removed the bird from a plastic bag and examined it. The eagle was in better condition than I had expected. Its face was bloody, and it had a large wound right in its belly, but the bird was completely intact, and the feathers looked decent. Adam extended the enormous wing and combed through the feathers. Explaining that each eagle has ten primary feathers and fourteen secondary feathers, he counted them all up and checked the quality of each one. The wing was looking good, and so was the other one, but when he got to the tail, there was a problem.

"Oooh, what happened there?" Atencio asked.

The tail was missing six out of its twelve feathers and would have to go. Adam took a huge red bolt-cutter-looking thing and chopped the tail off in one crunchy click. Then he took the tail and put it in the trash can.

Meanwhile, on the eagle-processing table opposite from where Adam was set up, longtime repository specialist Dennis Wilst was hard at work on another South Carolina bird, which was notable for its extremely well-preserved head.

"This guy is amazingly fresh," Wilst announced to the room. "He still has an eyeball—you don't often see that."

The eyeball might have been fresh (it didn't look that fresh to me, but what did I know?), but the head itself was a different story. Whoever had found the eagle had probably put it in a freezer right away—thus the fresh eyeball—but the ice in the freezer had soaked the head and turned it into a "big mess."

Wilst and Atencio discussed how to proceed. If they put the head into the freezer immediately, the eyeball would stay fresh, but the rest of the head would be compromised. On the other hand, if they tried to dry out the head first, the eyeball might suffer. After some back-and-forth, they decided to try drying out the head for a while. Wilst retrieved a pair of bolt cutters, chopped off the head, and put it aside. I, of course, snapped a picture. As for the rest of the bird, again the wings were fine, but the tail was, in Wilst's words, "not looking too good." This time, Wilst cut off the wings and put them aside to dry, disposing of the rest of the bird in a nearby trash can.

Honestly, I could have watched this all day, but I had stayed long enough and didn't want to wear out my welcome. I thanked everyone and took my leave. On my way out of the refuge, I dodged prairie dogs with my rented Hyundai and thought about what I had seen. The repository was better than nothing, that's for sure, but could it really be the ideal way to resolve the conflict between how the Native Americans want to practice their religion and how to best protect the environment? I wasn't convinced.

Native American tribes like the Hopis, the Navajos, and the Northern Arapahos have been taking bald and golden eagles from the wild for use in their sacred ceremonies for probably thousands of years. These tribes took their first eagles long before environmentalists embraced the eagle as a symbol in the mid- to late twentieth century, long before the Framers chose the eagle to be the new nation's symbol in the late eighteenth century, indeed long before Christopher Columbus "discovered" America in the late fifteenth century. Taking an eagle and using it, or its parts, in these sacred rituals, such as the Sun Dance, is a fundamental aspect of the identity of these Native Americans. Consider the comments submitted to a federal trial court by the Northern Arapaho Tribe in a recent case that I'll discuss later:

> Eagles, in the Arapaho culture, are revered and used for religious ceremonial purposes. . . . An Arapaho does not go out with the purpose to "hunt" an eagle. An eagle presents itself and donates its "holy body" to an Arapaho who needs it for ceremonial purposes such as with the Sun Dance. The eagle is a messenger to and from the Creator. . . . The eagle is sacred and has been part of Arapaho culture since time immemorial. . . . The Sun Dance is vital to the religion of the Arapaho people. Our

deep connection to eagles is a vital and necessary component for the cultural survival and religious identity of the Arapaho people. . . . The history of the NAT [Northern Arapaho Tribe] is filled with efforts by non-Indians to suppress NAT traditional religious practice. . . . Eagles which arrive at the Federal Repository have died by any number of means, some killed intentionally or knowingly, some killed by accident. . . . As a result, eagles supplied by the National Eagle Repository are inappropriate for the Sun Dance needs of the Arapaho people. . . . The take of an eagle is itself a religious ceremonial practice and is an important part of the larger Sun Dance ceremony.

Still, given the growth of environmental consciousness in the late twentieth and early twenty-first centuries, it's hardly surprising that the practice of taking eagles has become controversial. A few years ago, *Audubon* magazine ran a piece critical of the Hopi practice of taking eagles. Without a doubt, many people, particularly readers of the bird-focused publication, shared the views of the article's author, Ted Williams, who concluded:

When I asked [a Native American rights activist] if all Native Americans should be able to take wildlife from all park units, she responded with an emphatic "Yes." Then she said: "If you're exercising your religion, it doesn't matter what other people think about it." But it does matter. In America freedom of religious belief is absolute. Not so freedom of religious practice. Religious practice has always been questioned when it conflicts with the public good. . . . What about the rights of eagles? And what about the rights of Americans—white, black, and red; young, old, and yet unborn—who cherish or will cherish the sight of living eagles? . . . What kind of gods really want eagles dead instead of soaring in our spacious skies?

The article attracted a number of spirited responses from both sides. A married couple from Port Townsend, Washington, wrote:

"*Audubon* . . . will sit at the right hand of God for this defense of the golden eagle. . . . A long time before the Hopi Tribe existed, the Greeks thought of the eagle as messenger of the gods. Only fools kill the messenger. We cannot remain neutral on this issue."

On the other hand, the chief of staff of the Hopi Tribe's chairman took great umbrage with the piece, calling it "a disgrace and an insult." He said that Williams "makes no effort to present the importance of eagles in Hopi religion. Rather, [he] depicts the practice in a most derogative manner and dismisses Hopi beliefs as pious hocus-pocus."

To that, Williams responded: "I have enormous admiration for the Hopi. But they need to remember that in addition to being Native Americans, they are Americans. They're entitled to their religious beliefs, but in my humble opinion, it's time for them to modernize their religious practices."

Whether or not Williams is right that the Hopis should modernize their religious practices—if this book makes one point, it's that questions like this are really difficult—he is on the mark about the distinction between religious belief and practice, especially when it comes to the United States Supreme Court, which has struggled for well over a century with the question of whether the government can prohibit people from doing things they think are required by their religion.

The Court first faced this question in the late 1800s, when members of the Mormon Church challenged the government's criminal prohibition on polygamy. The plaintiffs argued that the government's ban infringed their rights under the First Amendment's Free Exercise Clause and demanded an exemption from the general prohibition. The Court disagreed, finding in *Reynolds v. United States* that while the government may not infringe on religious "belief and opinions," it could prohibit religious practices that are "subversive of good order."

That was how the law stood until the 1960s, when the Court took a sharp turn toward protecting religious freedom. Adell Sherbert was a Seventh-day Adventist who refused to work on Saturdays.

When she was fired by the textile mill where she worked and ap-plied for unemployment benefits, North Carolina refused on the grounds that Sherbert lacked "good cause" for refusing to work on Saturdays. In *Sherbert v. Verner*, the Court held that the government could not impose a substantial burden on religious practice with-out passing strict scrutiny and that North Carolina had not met its burden under this demanding standard. (As I explained in chapter 1, strict scrutiny is the most demanding test in constitutional law for the government to survive.) The state had to pay up. A few years later, the Supreme Court applied the *Sherbert* approach in a case involving the Amish in New Glarus, Wisconsin. Three Amish fami-lies refused, on religious grounds, to send their sons to school after the age of fourteen, even though the state required all children to attend school until they turned sixteen. When the state fined the families five dollars, the Amish argued that the fine violated their religious freedom rights. In the case of *Wisconsin v. Yoder* (possibly the zenith of religious freedom law in the history of the nation), the Court, in a nearly unanimous decision, agreed with the Amish.

Things changed again in 1990, when the Court heard a case in-volving two members of the Native American Church (a peyote-based Native American religion that has taken on some elements of Christianity). The men claimed that Oregon had violated their free exercise rights when it rejected their claims for unemployment compensation on the grounds that they had been fired for ingesting peyote. The parties in the case all agreed that under *Sherbert* and *Yoder*, the relevant legal standard was strict scrutiny and briefed the case accordingly. The Court, however, in an opinion written by conservative loudmouth Antonin Scalia (and joined by the other-wise lovable liberal John Paul Stevens), decided that strict scrutiny would no longer be the relevant standard for evaluating generally applicable, neutral laws that happen to burden religious practice. In other words, the Court took its cue not from the recent *Sherbert* or *Yoder* cases but rather from the *Reynolds* polygamy decision that the Court had handed down 111 years earlier, when the country had only thirty-eight states and Rutherford Hayes was president. Under the new rule of *Unemployment Division v. Smith*, the government is free under the Free Exercise Clause to place substantial burdens on religious freedom through generally applicable criminal and civil laws. No longer would religious believers be entitled to exemptions from such laws under the Constitution because, as Scalia put it, "any society adopting such a system would be courting anarchy." Anarchy, yikes!

At the time the ruling came out, the *Smith* decision was one of the most unpopular decisions the Supreme Court had ever issued. It was surely the only decision that attracted virulent criticism from both the American Civil Liberties Union and the Christian Coalition. In the wake of the ruling, not only did Oregon pass a regulation exempting religious use of peyote from its general drug laws, but religious and civil liberties groups from all over the political spectrum also came together to pass the Religious Freedom Restoration Act or RFRA. The statute restored the strict-scrutiny standard of *Sherbert* and *Yoder* to any law imposing a "substantial burden" on religion. A few years after RFRA's enactment, however, the Supreme Court struck the law down as it applied to states and local governments on arcane constitutional grounds having to do with limits on congressional power. Congress responded by passing the Religious Land Use and Institutional Persons Act (RLUIPA), which I discussed earlier, to at least restore strict scrutiny for burdens imposed on religions by state and local correctional institutions and zoning laws. It is pretty clear, for other arcane constitutional reasons, that RLUIPA can be applied in these contexts. Moreover, although RFRA does not apply to the states or local governments, it continues to apply to the federal government. Indeed, RFRA is the statute that the Court applied in the (in)famous *Hobby Lobby* ruling that the federal government cannot force closely held companies to provide contraception coverage to their employees if the owners object to such coverage on religious grounds.

Because the Bald and Golden Eagle Protection Act is a federal law, most challenges brought to its limitations by Native Americans have involved RFRA. The most prominent challenge is the case of *Friday v. United States*, which was decided by the Tenth Circuit Court of Appeals in 2008. When Wilson Friday, a Northern Arapaho man, shot and killed a bald eagle for use in the tribe's Sun Dance ritual, federal prosecutors charged him with violating the eagle protection act. In response, Friday argued that the prosecution violated his rights under RFRA by substantially burdening his religious practice. The feds defended the prosecution on the grounds that the prohibition on taking eagles furthered the government's compelling interest in protecting bald eagles. Moreover, the government argued that the prohibition was narrowly tailored, as required by RFRA, because Friday could have applied for a live-take permit from the Fish and Wildlife Service but didn't. (Friday could also have theoretically applied to get an eagle from the repository, but his position was that an "impure" eagle from Commerce City would not suffice for the Sun Dance, and the government conceded the point.) In response to the argument that he could have applied for a permit, Friday argued that he didn't even know about the permit system and that, in any event, applying for a permit would have been futile, since the government has only ever granted a handful of them. The trial court held for Friday, opining that the permit process was "biased and protracted" and concluding that "although the Government professes respect and accommodation of the religious practices of Native Americans, its actions show callous indifference to such practices."

Unluckily for Friday, however, the appellate court sitting in Denver reversed the trial court and held for the government. In a unanimous opinion written by super-pro-religious-freedom judge Michael McConnell, who as a law professor at the University of Chicago had been the academy's most outspoken critic of the Court's *Smith* decision, the appeals court held that the permit system was adequate to balance the government's interest in protecting eagles and Friday's right to practice his religion. McConnell concluded that even though (1) Friday didn't know about the permit system, (2) FWS didn't publicize the permit system, and (3) few people had ever used the permit system, Friday still could have and should have used it. McConnell left open the possibility, however, that if the permit system does not in fact work as it should, the court would be open to considering another suit in the future: "We are not oblivious to the possibility that the government's permit process for the religious taking of eagles may be more accommodating on paper than it is in practice. If so—if the process is improperly restrictive, burdensome, unresponsive, or slow—we trust that members of the tribe will not hesitate to vindicate their rights either through petition or in a proper suit."

So, then, how has the permit system worked out for the Northern Arapahos? In 2009, the tribe applied to FWS for a permit to take eagles within the Wind River Reservation in western Wyoming. Ten months later, the agency told the tribe that its application was incomplete. The tribe then submitted a complete application, asking for permission to take one or two adult bald eagles every year on the reservation. The permit application was complicated by the fact that the Arapahos share the reservation with the Eastern Shoshone Tribe, which does not share the Arapahos' belief that live eagles must be taken for religious rituals and which therefore opposed the permit application. FWS considered the petition for about eighteen months and then granted permission for the Arapahos to take two bald eagles every year in the state of Wyoming, but *not* within the reservation.

At first, the decision was a comical result, since Wyoming also has a law prohibiting the killing of eagles, but even after Wyoming agreed that the Arapahos could take two eagles within the state, they were not satisfied, because they believe the eagles must be taken within tribal lands. The tribe sued, claiming that FWS's refusal to grant it a permit to take eagles on the reservation violated its rights under the First Amendment and RFRA. In late 2012, the trial court rejected the tribe's RFRA claims, but the Arapahos persisted in pressing their claims under the First Amendment. In March 2015, the trial court held that the FWS had in fact violated the First Amendment by refusing to grant the permit. The FWS has appealed the trial court's decision to the federal appeals court, which could potentially take years to reach a decision in the case. It seems pretty clear that the government's permit process was just a little bit more "accommodating on paper" that it was "in practice."

Gracie was having a good day. This was somewhat unusual. On most days—especially those when noisy construction was going on outside—she liked to hang out alone in a corner of the football-field-length shelter she shared with a dozen or so other birds, enjoying looking at herself in a little mirror perhaps, but not doing much else. Not that anyone could really blame her. Since being hit by a car, she has been blind in one eye and has so much shoulder pain that she often kept her head almost upside down to get some relief. But on this day, Gracie was out of her corner, hopping sideways like a crab, staring me down with her one good eye as I crooned, "Hello, Gracie," to her over and over like a fool.

Gracie is a young bald eagle, one of forty-plus injured eagles that live in the Iowa Tribe's Grey Snow Eagle House in Perkins, Oklahoma. About a dozen of the birds are golden eagles; the rest are bald eagles, though only some of the so-called balds are old enough to sport the classic white head that gives them their name. The birds have all been rescued from something or another—a few had flown into electric wires, a couple had broken bones that fused together improperly, one had been injured by a wind turbine. A good number are from Oklahoma, but some birds came from as far away as New Mexico or Oregon or even Connecticut. A few—those who have two eyes, two working feet, and can learn to fly again—will recover enough to be released back into the wild, but for most, Grey Snow will be their home for life. In captivity, by the way, bald eagles can live for over fifty years.

Apart from the macabre eagle repository and the Kafkaesque permit system, the other way that Native Americans can legally get their hands on eagle feathers (though not whole eagles or other parts) is by operating a tribal aviary for injured eagles. The FWS has granted permits to a handful of tribes—the Zunis, for example, and of course the Iowas—to run these aviaries and collect molted feathers for distribution to tribal members. Ever since I had visited the repository, I had wanted to make a trip to one of these aviaries because I thought that witnessing the feel-good stories of people helping injured but still beautiful birds would nicely counterbalance the visions of dead, bloodied, and decapitated eagles from Commerce City—visions that had been stuck in my head for months. I was right.

When I arrived at the aviary, I was a bit shaken up because I had just driven some sixty miles from Oklahoma City to Perkins through what might have been the most insane rainstorm I've ever seen in my life. The radio had said that there was a 50 percent chance of seeing a tornado within a hundred-mile radius of the city, and although I don't think I saw one, this might have been because for most of the drive, I was unable to see farther than about three feet in front of my car's hood. In any event, by the time my guide Amy started showing me around the place, the rain had stopped and the sun had come out, which made it far more pleasant to stroll around the grounds and say hello to all the magnificent birds that call the aviary home.

Before we looked around, though, Amy gave me some background. She was a recent zoology grad from Oklahoma State who had been working this quirky job for two years (among other things, she was in charge of the incredibly stinky building where the aviary keeps the hundreds of rats and rabbits that the eagles get to eat). The aviary, she said, was first opened in 2006 with only four birds. The main purpose of Grey Snow is to help the injured eagles, with the provision of feathers to members of the tribe being a side benefit. The staff also educates the public about eagles and Iowa Tribe of Oklahoma culture by offering public tours on the weekend and inviting school groups to visit during the week. Unfortunately, however, when I was taking my tour, Harley, the guy in charge of the aviary's education program, was busy out in the garage, chopping up two freshly killed deer as a treat for the winged residents.

Being kind of obsessed with feathers, I asked how they were collected, and Amy taught me a couple of interesting things. Eagles apparently molt exactly half their feathers each year, in a precisely symmetrical fashion. If an eagle molts her third wing feather on her left side, she will also molt the mirror image third wing feather on the right. Having even sets of feathers, it turns out, is pretty critical when you are an eagle diving at two hundred miles an hour toward your prey. I also learned that in Iowa culture, women are not allowed to touch a feather unless it is given to them; for that reason, whenever Amy touched a feather, she did so wearing gloves (even though she herself, unlike the others who worked at the aviary, was not a member of the tribe).

Our first stop on the tour was the so-called flying cage, an enormous structure where nine bald eagles were relaxing and frolicking and squawking up a storm. The cage was made from wooden slats with narrow openings to the outside for fresh air and other elements to enter. The eagles apparently love water, and so they were in a particularly good mood given the recent rain that was still dripping on them. I was introduced to Scrapper, another car crash victim brought back from the "brink of death" by the aviary's veterinarian in Tulsa (who donates all his time to Grey Snow for free). Scrapper was now sitting quietly on a high-up perch. I also met Sally, who had lived through a complicated bone surgery that apparently has a mere 20 percent survival rate.

Our next stop was the intensive care unit, which at the time was merely a small room next to the main office (a new, larger ICU has since been built). This is where the staff cares for the acutely injured birds or those with a temporary condition requiring special treatment. The day I was there, the ICU's sole resident was Newman, a female bird who had been named somewhat prematurely and was now suffering from a "bumble" on her toe, which is something that is apparently really painful and gross. I bent down to Newman's cage and introduced myself. She was fairly sedate, which Amy explained was a new state of affairs that had started when the staff put some dark blue padding around the cage to stop her from slamming herself against the wood frame. Whether Newman's change in attitude could be chalked up to the padding itself or the dark blue color, however, remained a mystery.

Gracie made her home in the "handicapped" cage, which was much like the flying cage, except that the residents cannot fly, or if they can fly, they can only fly a little. Here the perches were much lower, and many were connected to ramps so the hobbled eagles could hop up more easily. A lot of the birds in this cage had amputations, like Nubby, who had an amputated wrist. Then there was poor Banana, who came to the aviary with a leg deficiency that made it almost impossible for him to walk, much less fly. Some veterinarians had suggested that Banana be euthanized, but the aviary's vet engineered some kind of eagle hammock that helped him recover to the point where the staff was able to train him to walk at least short distances. When I was there, Banana was playing with a small plastic orange egg. He seemed mildly happy. Or at least not particularly unhappy.

The handicapped cage was, at the same time, both the saddest and most joyful part of the aviary. On the one hand, here were birds who had suffered dearly at the hands of power wires, wayward vehicles, bad luck, and all sorts of other things, but on the other hand, the eagles were still very much alive, which they would certainly not have been were it not for the aviary. And not only were the eagles alive, but they were also molting feathers, which members of the tribe could collect to use for religious purposes. In many ways, this was an ideal approach to the problem of eagle feathers—no birds had to die, and nobody had to file forms with the federal government to receive any religious items. All told, the aviary was a wonderful place. Amy told me that it is even considered a holy place by many members of the tribe, some of whom come to the aviary from time to time to smoke a sacred pipe, pray, or otherwise engage in religious activities. And when an eagle dies, the tribe gives it a religious burial rather than sending it over to the repository to be cut up and distributed.

On the other hand, great as they are, tribal aviaries cannot be the entire solution to the eagle problem. Not only are they difficult and expensive to build and operate, but there are only so many injured eagles that can take up residence at such places. Moreover, because eagles molt relatively infrequently, it would be highly unlikely that even, say, double the number of aviaries that exist now could supply enough feathers to meet the demand. Finally, of course, many Native Americans want more than just the feathers. As the Arapaho case demonstrates, for some Native Americans, nothing short of a whole eagle, taken in the wild, will suffice. Aviaries, then, can and should be one important method of addressing the controversy over the religious use of eagles and their parts, but difficulties will inevitably remain. We still have to face the question of how to balance both our interest in preserving these beautiful birds and the tribe members' interest in freely practicing their religion. It's a difficult problem and one that leads into the larger issues of the clash between religious practice and environmentalism. I kept on traveling to see if I could find some answers.

5

SINGAPORE SMOKE

*Technological Fixes
and the Impacts of Regulation*

At about three in the afternoon on the day I arrived in Singapore, I found myself lurking behind a pillar on the corner of the city's bustling Chinatown neighborhood, trying to stay inconspicuous. Across the narrow street in one direction was a tiny Taoist temple with a metal burner about the size of a hot dog cart parked outside the entrance. I watched as a lady with a pink shirt slowly and methodically fed the burner with sheets of joss paper—a type of paper, often printed to look like money, that is burned at certain times of the year by Taoists to appease the ghosts of dead ancestors. Across the street in the other direction, a group of men and women had gathered around a large open pen, maybe eight feet long, five feet wide, and two feet tall, and were filling it completely with joss paper of all different colors and sizes. At first I thought that perhaps they were planning to sell the paper to worshippers who would then burn it at the temple. It soon became clear, however, that they were in fact going to fill the entire pen with joss and then set it on fire. I was delighted!

I had come to East Asia and Southeast Asia—five days in Singapore, followed by a few days in Hong Kong and then a week in

Taiwan—to investigate the religious practice of burning joss paper and its effect on the environment in the region. Though August was hardly the most comfortable time to make the trip, I went then because that is when the Hungry Ghost festival falls—the month of the year (specifically, the seventh lunar month in the Chinese calendar) when Taoists believe the ghosts of everyone's ancestors are released temporarily from hell so they can come visit earth and catch up on what's happening here. The burning of joss sheets and all sorts of other items made out of paper (shirts, cigarette cartons, watches, even paper iPods) is particularly pervasive during Hungry Ghost month; believers burn this stuff as a way of pleasing and appeasing the ghosts. During my time in Asia, I wanted to understand a number of things—how much air pollution the burning causes, what these societies and legal systems have tried to do about the problem, and how these efforts have affected the traditional ways of practicing religion in the area. At the same time, I also wanted to see just how humungous a bonfire I could actually witness in the limited time I was there. And here I was, only hours into my trip, and it looked like I was already going to see an enormous open conflagration right on the corner of a busy neighborhood.

I waited and watched patiently, as the group of twelve or so worshippers piled their paper into the pen. They put the types and colors of paper into the pen in a particular order and then added the finishing touch by draping red streamers embossed with golden Chinese characters over the top of the pile. The pen was completely filled and ready for burning. I rubbed my hands together in eager anticipation and waited for the imminent religious ceremony to begin—maybe some chanting, or bowing, or recitation of prayers—followed by the lighting of the whole giant heap on fire.

Instead, everybody left. "Hey, hey, where are you going?" I wanted to call out after them. But no, they were definitely gone. I wasn't sure what to do. I could have camped out on the corner, waiting for someone to come back and burn the paper, but what if that took eight hours? Or two days? Plus, I was jet-lagged. I needed a nap. So I went over to the neighboring hawker center—the awesome Singaporean version of an outdoor food court, where for very little money you can get everything from fish-head curry soup to Chinese noodle dishes to fried bananas—and asked an old lady if someone was going to come back and burn the pen of paper. We conversed for a bit, mostly in English, with some leftover college Mandarin and hand signals mixed in. I learned that at seven thirty that evening, there was going to be a big to-do involving singing, praying, and burning. I was psyched. I returned to the hotel, took my much-needed nap, and at about seven o'clock, headed back to see my Day One Joss Jackpot.

Of course, though, when I returned to the corner, the pen was empty. Some ash residue was left on the street under and near the pen, but there wasn't so much as a single piece of paper left, where before there had been thousands and thousands. I had completely missed the fire.

As it turned out, a Hungry Ghost ceremony did take place at seven thirty; it just did not involve burning, or at least that much burning (there was a little burning). A temporary stage had been set up on the street with lines of folding chairs in front, and a swelling crowd was arriving and taking their seats. I had heard of these things. Known as *getai*, they are shows, usually involving music, that are put on to entertain the ghosts and keep them happy while the spirits roam the earth for the month. The first row of seats is kept empty because that is where the ghosts are supposed to sit. I walked across the street and sat on the second floor of the hawker center, where I would have a bird's-eye view of the festivities.

Let me tell you, if this show was our best hope to keep the hungry ghosts from wreaking havoc on our world, then we are in pretty bad shape. The show started off okay as a live band ripped into an instrumental version of the 1970s Donna Summer hit "Hot Stuff," which I thought was an appropriate selection for a hell-related festival. Unfortunately, this was the high point of the evening; I sat upstairs watching for about an hour, hoping that maybe at some point someone would burn something big, as a variety of kitschy singers in vaguely space-age outfits came up to the stage and sang cheesy tunes backed by an awful electronic beat. It reminded me a lot of the bar mitzvahs I regularly attended in my hometown in the northern suburbs of Boston back in 1981, only without all the amateur French kissing. After a while, I couldn't take it anymore, so I got a beer and headed back to the hotel to call it a night. My Joss Jackpot would have to wait for another day.

Singapore is an interesting—some would surely say bizarre—place. A tiny city-state located at the southeastern tip of the Malaysian peninsula that extends down from Thailand into the South China Sea, Singapore is often described as a red dot in a sea of green, because it is a largely Chinese city surrounded by the much larger Muslim countries of Malaysia and Indonesia. In fact, the city is incredibly diverse. More than 70 percent of the population is Chinese, but there are significant numbers of Muslim Malays, Indians, and Westerners, a holdover from Singapore's history as an English colony. On any given day while riding Singapore's sleek modern subway, you might see (hypothetically) an Asian guy with red-tinted hair, orange shorts, and a purple shirt standing next to a pudgy Muslim woman with a red head covering and a black and white print dress, sitting across from an Indian guy with a pink shirt that says OPIUM on it, talking on the phone and looking at a beautiful young woman of unidentifiable identity, who is keeping her distance from a dumpy white guy taking notes for a book about religious practices that harm the environment. It is this diversity that makes Singapore so unique and fascinating. Most public signs are written in four languages (Chinese, English, Tamil, and Arabic), and many people speak both Chinese and English, the language that is primarily used in schools, although it is a quirky version of English and some other languages and is often referred to as Singlish.

The diversity of Singapore directly affects the issue of joss burning in a way that you do not see in less diverse places like Hong Kong or Taiwan. Joss gets in people's faces. Literally. Whether it is burned in a bin outside someone's apartment block, on the side of the street in front of a business, in a pen on a crowded corner while I'm taking a nap, or in a burner in a temple, or anywhere else as densely populated as the places I was visiting, the smoke is going to get into other people's eyes, throats, and lungs. This is bad enough when the population is relatively homogeneous, but in Singapore, where you might find a Hindu temple, a mosque, a church, and a Taoist shrine all within the same block, tensions over joss smoke can get extra fiery.

The Internet is filled with stories and comments sections illustrating this battle of the cultures. In 2012, a Muslim actor caused a brouhaha when he attacked the practice of joss burning on his Facebook page: "If you think our prayer's call is kind of noise pollution, what about those 'burning activities' you have out there? Talking about Global Warming? I think you're just Gorblok." (*Gorblok* is a Singlish word for a fool or an idiot.) The actor took down the comments after they attracted public criticism. Here is a fairly typical comment on an online forum from the year before: "I'm Catholic and I don't mind it unless they burn big black patches in fields or be dumbasses and decide to burn a HUGE pile at the same time causing flying burning offerings to burn down other things." Another Christian commenter compared joss burning to the smog from Indonesian forest fires that practically shut down Singapore in the summer of 2013; a response urged the commenter to "be a patient Christian" and "have some religion [*sic*] tolerance." Another person, apparently lacking such tolerance, posted pictures of a burned-out field in front of his house and proclaimed: "Seriously wtf is wrong with these idiots?" And in August 2011, a man died of a heart attack after two sisters threw a bunch of flowerpots at him because he was burning joss paper and incense outside his home.

I wanted to understand more about the racial dimension of the tension over joss, so I made an appointment to meet with National University of Singapore professor Tong Chee Kiong, perhaps Singapore's leading expert on ethnic and religious diversity in the country. We met at something called the Kent Ridge Guild House, a club for NUS alumni and faculty. Tong drank glass after glass from the private bottle of whiskey he keeps in the club, while I drank Tiger beer and wondered why a Scorpions album was playing on the club's sound system. Luckily, I had checked the club's website earlier in the day and learned that the guild's dress code for week-days was "smart casual," so I was wearing some "closed shoes" and "proper trousers." I shudder to think what would have happened if I hadn't seen the website and had shown up in the purple clown pants I had originally chosen for the afternoon.

Among other interesting things, Tong described how environmental laws have significantly affected the practice of joss burning in Singapore. It used to be that Taoist families would primarily burn joss inside their homes, a preferable practice because it ensured that the ghost ancestors would know who was making the offering and to whom it was intended. When the government prohibited open burning within the home, families instead burned on the ground outside their homes. But since this too is dangerous, the government started providing bins—they look just like trash cans—for the apartment complexes, where most people in Singapore live. The residents were required to burn their joss inside the bins.

The effect of these laws in Singapore has been to turn a private family religious ritual into a public communal one. Although some families and individuals still burn on the ground or in their own container (once I even saw a couple of guys burning paper in a wok) outside their homes or businesses—many either use the bins or burn joss at temples. On the whole, Tong suggested, the Chinese worshippers do not really like this communal practice so much, because if everybody is burning their offerings together, how will the ghosts know who the offerings are for? Still, the Taoists in Singapore have largely been willing to follow the laws. Things would be different, however, if the government actually tried to prohibit burning altogether. This, Tong said, would be over the line. Indeed, when one local official tried to ban joss burning, Lee Kuan Yew—the prime minister at the time and the most important and famous

political figure in Singaporean history—had to personally intervene to avoid political disaster for his party.

Joss burning may not be the biggest source of air pollution in places
like Singapore and Hong Kong—far from it—but it is still a big deal.
It is unclear exactly how much joss is burned in Singapore, but
data from the Taiwanese Environmental Protection Administration
suggests that close to three hundred thousand tons of joss paper
is burned every year in that small island nation. And the stuff is
terrible for the environment, particularly for the local area around
where the joss is burned. For whatever reason, scientists in Taiwan
have been doing the most research on the effects of burning joss,
and their studies clearly show the harmful effects of the practice
on the environment and human health. In scientific papers with
titles like "Polycyclic Aromatic Hydrocarbon Emissions from Joss
Paper Furnaces," "Characterization of Polycyclic Aromatic Hydrocarbon Emission from Open Burning of Joss Paper," and "Removal
of Particulates from Emissions of Joss Paper Furnaces" in fun-filled
journals like *Atmospheric Environment, Journal of Hazardous Materials*, and *Aerosol and Air Quality Research*, these scientists have
demonstrated that joss burning significantly increases the amount
of particulate matter, metals, polycyclic aromatic hydrocarbons,
and other pollutants in the ambient air, all of which can cause cancer and serious respiratory ailments.

After my conversation with Professor Tong, I spent a couple of
days temple-hopping to see what I could find out about how joss
burning really takes place and how it affects the air inside and
around the temples. My first stop was a place called the Lian Shan
Shuang Lin Temple and Monastery, pretty far out from the center in
Toa Payoh, which is considered a heartland town, meaning that it's
the kind of place where most people in Singapore live, generally in
enormous housing complexes administered by the government. In
English, the name of the temple translates into something like "The

Twin Grove of the Lotus Mountain Temple." Lian Shan Shuang Lin was founded by a father and son in the late nineteenth century as a place to stay for some monks who were on their way back to China from a trip in India. Like many of the temples I visited in Singapore and Hong Kong, this one was originally built in a desolate area, but as the population blossomed, more and more people started moving in all around it. The spacious temple and monastery grounds now sit right in the middle of a packed residential area; a walk from the temple to the nearest high-rise apartment complex takes only a couple of minutes.

Most of the temple and monastery is for Buddhist worship and is smoke-free except for some incense burning here and there, but one temple building in the front, where the taxi left me off, is completely different from the rest. This is the temple devoted not to any Buddha, but rather to the town god. A newspaper article posted at the entrance explained that because the town god administers justice in hell and supervises the return of ghosts to hell at the end of the Hungry Ghost festival, the temple's architecture is not tall or spacious, and the place exhibits "an air of gloom and eeriness reminiscent of the nether world." They weren't kidding. Directly outside the entrance of this little temple stood two giant joss-burning furnaces shaped like ornate pagodas. I sat and watched for about an hour as person after person after person came with stacks of paper and loaded them into the hot fire inside the furnaces. Smoke and ash and little scraps of crispy paper belched out of holes in the top of the fifteen-foot structures, turning the whole place into something resembling the inside of a barbecue grill. By the end of my time there, my eyes were red and itchy, and I smelled as if I had just gone swimming in a giant ashtray.

Next, I took the metro out to a temple called Phoh Kiu Siang T'ng. This was a very different kind of place from the first temple. It was in the middle of a residential neighborhood (not a high-rise government complex) and tiny. An article in the Singapore newspaper the *Straits Times* had piqued my interest in the temple when it reported that, just a month earlier, the place had purchased and installed two state-of-the-art burners that emit almost no smoke at all for a seemingly exorbitant amount of money. Apparently, the temple installed the burners in response to increasing complaints from nearby residents about the smoke that was emitted by the previous burners. According to the temple spokesman, "Some of the residents have asthma, and are very vocal during the lunar seventh month, to the point where even the police are called in."

Even though one of my goals in coming to Asia was trying to see the most nasty paper-burning event possible, I also had my eye out for environmentally friendly burning situations, so I could compare the good with the bad. The burners at this temple turned out to be a perfect example of what I was looking for. It was easy to see how these twenty-foot-high golden pagoda-shaped burners would be good for the environment. Fitted over the vents that would have ordinarily sent the smoke from the burning joss paper into the air were thick silver-colored pipes that rerouted the smoke into a large, unsightly filter located directly behind the burners. The filter presumably gets rid of all or most of the smoke, although I have to admit that I do not know for sure how much of the smoke is removed by the filter, because in the hour or so that I awkwardly stood around waiting to find out, not a single person came to the temple to burn any paper. Unlike the bustling traditional temple I had just visited hours before, this place was a veritable (excuse the pun) ghost town. Even when I came back a second time—ten days later, on a weekend afternoon, on my way back through Singapore before returning to the United States—I didn't see anyone at all using the burners.

There are, of course, all sorts of possible reasons why I didn't see anyone burning joss paper in my two trips to visit these fancy-pants pagodas. As a small, local temple, it is not frequented by all that many people in the first place. Perhaps people tend to come at night. Or maybe I was just unlucky. But what kept occurring to me as I sat, bored, watching the unused burners, was that maybe, just maybe, people don't really want to burn joss paper in a burner where the smoke goes into a pipe and into a filter and then comes out, if at all, as a tiny puff of purified air. Doesn't taking away the smoke kind of take the oomph out of the whole experience? What a bore, to burn a bunch of paper and have nothing actually come out of the burner. Sitting there looking at the lonely pagodas, the structures ornately decorated with carved pictures of deities and gleaming under the hot Singapore sun, I thought about how encouraging or requiring religious practices to be more environmentally friendly can run the risk of making religion dull and pedestrian, perhaps even discouraging people from practicing their rituals. Is protecting the environment, I wondered, always worth the cost?

The following day, I visited a third temple, a place that was yet again very different from the ones I'd seen so far. The Kong Meng San Phor Kark See Monastery (also called the Bright Hill Temple) is the biggest Buddhist temple in Singapore. Probably a hundred Phoh Kiu Siang T'ngs could fit inside it. This third temple was far from the city center; to get there, I had to take a bus that sported a sign announcing NO DURIANS ALLOWED. The sign referred to the spiky, odiferous national fruit that nobody wants to sit next to in a closed, crowded space.

Although Singaporean Buddhists are not big joss burners, I was visiting a Buddhist temple because Dr. Tong had explained to me that in Singapore, many Taoists bury their dead ancestors in Buddhist temples and burn joss for them there. Apparently, the Taoists like the burying facilities at Buddhist temples and prefer to bury their dead in a place where there's a lot of chanting and praying going on, even if it's not quite the right chanting and praying. The Buddhists allow the Taoists to bury their dead in the Buddhist temples because of some combination of being compassionate and wanting the money that they charge for burial. But the Taoists still want to burn joss paper at certain times during the year at the site where their dead ancestors are buried, so the Buddhist temples will set up an area, maybe in the back of the temple, where the Taoists can burn their paper to honor their ancestors. In other words, the temple is predominantly Buddhist, but there is one small area that the Taoists use for burning.

I thought this arrangement was extremely weird when Tong explained it to me, but sure enough, that's exactly what I found at Bright Hill. At the far back of the grounds of this giant temple, past the enormous halls for Buddhist worship, past the bodhi tree and the koi ponds, past the Pagoda of Ten Thousand Buddhas and the Hall of Universal Brightness, you finally come upon the Pu An Columbarium. It is here at this bleak furnace that Taoists come bearing stacks of joss paper, which they hand over to a temple worker in a yellow uniform. The worker then throws it all into the furnace, where it's burned and turned into smoke that comes belching out the top of a giant metal apparatus and into the air around the temple and surrounding neighborhood. Of all the places I visited to watch joss being burned in my travels, this was one of the most depressing. Two big English-language signs had been hung in front of the furnace urging people to PROTECT THE ENVIRONMENT by BURNING LESS JOSS, but the signs were so blackened from the ubiquitous soot that they were nearly impossible to read.

Luckily, I didn't have much time to sit there breathing pounds of polycyclic aromatic hydrocarbons into my increasingly abused lungs, because I had an appointment to meet with Nancy Tan, the temple's public affairs officer, to talk about some of the changes that the temple was making to their burning process in response to complaints from people in the neighborhood. Nancy is so nice and sweet that while we were talking in a lovely conference room, I actually wrote "Nancy is so nice and sweet" in my little notebook. I asked Nancy about the temple's relationship with the residents of the nearby neighborhoods, and she explained that up until a few years ago, the burner had been completely uncovered, so the ashes from the joss paper went everywhere. People who lived in the area were not happy. Since then, though, the temple has "worked hand in hand" with the government and the nearby residents to solve the problem. Recently, for example, the temple upgraded its burning facility, hired a team of employees trained in reducing ash emissions to take the paper from the worshipper and put it in the furnace, and began letting the furnace rest for a while after it gets too hot, an innovation that results in far less ash being emitted into the air.

I arrived in Hong Kong the day after a typhoon had passed through and shut the city down. Light rain was still falling, and the air was so damp that it took only ten minutes of walking through the crowded, haphazard streets until a midsize swamp developed in my slacks. Before I went to law school, I had briefly worked in mainland China for a law firm headquartered in Hong Kong and had visited the tiny, bustling city-state many times. I always loved it—the chaos, the endless streams of people, the bright, blinking neon lights, the mix of old and new, of things Western and Chinese. In the nineteen years since I lived there for six weeks in a crowded apartment with a coworker and an entire Chinese family, though, I had forgotten just how New York–ish the place is. Going from sedate Singapore to frenzied Hong Kong is like putting on an AC/DC song after listening to hours of classical guitar. The city is also crazy expensive. One night, as I sat in a hotel bar at the edge of Kowloon staring through two-story-high glass windows at the harbor, drinking margaritas, and watching the boats traveling every which way in front of the gorgeous, twinkling skyline of downtown Hong Kong Island, I accidentally spent forty dollars on a cheeseburger.

I only had two full days before heading off to Taiwan, so I packed the time I had in Hong Kong full of appointments and activities. The first day, I would talk to people; the second, I would tour the temples. My first meeting was with Dr. Tong Wai Hop, the head education officer at the Taoist Association of Hong Kong. In Hong Kong, the government actively regulates joss burning in temples and other institutional settings by requiring the installation of certain smoke-scrubbing technology (more on this later). I wanted to talk with a Taoist representative to find out how Taoists feel about this regulation of their religious practice.

As I took the subway out to meet Dr. Tong, I wondered what a Taoist office might look like. Would it be floating in the air, in the middle of park made out of marshmallows? Maybe it would be totally empty, except for thousands of butterflies. When I exited the subway into a particularly gray and grimy part of the city and found the office in a characterless building that also housed the offices of the Top Ten Snooker Club, however, I realized that my fantasies were misplaced. Other than a small prayer room with a little shrine and a red rug, the Taoist office turned out to be pretty much like any other office—a reception area, a few cubicles, and a meeting room, where I took a seat in an ornate wooden chair across from Dr. Tong and his young, stylish assistant, Sam. I am not sure exactly what I had thought a Taoist sage might look like—maybe something involving bare feet and a long, white beard—but Dr. Tong was wearing brown pants, a brown-plaid shirt, white socks, and navy-blue Crocs, a weird outfit for sure and not what I had expected. The interview got off to a slow start. The Taoists seemed skeptical of me, which I could fully understand, and the combination of the warm room and my anxiety about the interview was making me sweat like Richard Nixon drinking hot tea in a sauna.

As the conversation got going, though, we all relaxed a bit, and I was eventually able to learn some interesting things from Dr. Tong, who made it clear that he was speaking only for himself, rather than for all Taoists. When I asked him about the regulation of furnaces at temples, his irritation with the government was palpable. It was not that reducing the amount of smoke would make it hard for Taoists to worship the ancestors that made him angry. Indeed, when I mentioned the smokeless Singapore temple, he said, "It does not matter if the smoke is coming out—we believe the ghosts will receive it."

Rather, he was upset about the significant cost of installing and running the control technology, particularly because that technology has to be working all day long, even if worshippers are only burning paper for fifteen minutes. Tong said that the Taoists would prefer to use the money to directly help people by providing health care to the elderly and education and scholarships to young people than to achieve some "vague" amount of environmental protection. He stressed that the pollution created by burning joss is minimal, particularly in comparison with the pollution caused by industry. Although he said that he believes people ought to be urged to reduce their energy use and pollution, the government should not enforce such reductions through regulation because, among other things, this discriminates against his religion. "I don't think the government understands the Taoist perspective," he said. "We have had these habits for thousands of years. Why should we change them? This is discrimination." Altogether we probably only talked for an hour, but I feel as though I at least got some insight into how Taoists feel about the regulation of joss burning, as well as what kind of plastic shoes Taoists prefer these days. Before I left, I asked Sam and Dr. Tong if they could recommend some places I should visit to see how Taoists burn joss in Hong Kong (their suggestions turned out to be terrific, as we'll see later), and then I took my leave.

After a brief visit with Kin-Fai Ho, the air pollution scientist whose work on incense I discussed in chapter 1, I arrived at the Environmental Protection Department for my final interview of the day. I was met by Danny Lo and Duncan Wong, two terrific, funny, eager government officers who showed me into a cramped meeting room dominated by a cruddy yellow table. Not only were Danny and Duncan smart and fun to talk to, but they were also able to clearly explain exactly how Hong Kong law tries to deal with the dangers of joss burning in a variety of contexts. The key legal provision that the government employs is the Air Pollution Control Ordinance, a statute that was first passed in 1983 and subsequently amended several times to control different types of air pollution. Section 10 of the statute provides for the following actions:

> Where the Authority or an authorized officer is satisfied that the emission of air pollutants from a polluting process is causing or contributing to air pollution . . . the Authority . . . may give an air

pollution abatement notice . . . to the owner of the premises or to the person carrying out the activity requiring him:

1. to cease the emission of air pollutants from the premises or to cease the operation of the polluting process;
2. to reduce the emission of air pollutants from the premises or polluting process;
3. to take other steps to abate the emission of air pollutants from the premises or polluting process.

Anyone who fails to comply with a government-issued abatement notice can be fined a lot of money and even be sentenced to a prison term. According to Danny and Duncan, the government uses this provision to regulate temples and columbaria, and other permanent, significant sources of joss-paper emission. Because the resources of the office are limited, however, and because the people who work there have a lot of other things to do, the government does not necessarily regulate every single emission source. Rather, the agency is, as the guys put it, "complaint oriented." If a group of neighbors brings to the agency's attention some temple that is emitting a bunch of harmful smoke, then the government will do something about it. If nobody complains, however, the government won't even know about the emissions, much less regulate them.

I asked Danny and Duncan what the government requires of temples that are burning a lot of joss. Does it ban emissions altogether? Does it require the temples to limit their emissions to a certain level, and if so, to what level? The officers explained that instead of banning emissions or even requiring them to fall below some threshold amount, the government insists that the temples install what's known in the environmental-law game as best available technology (BAT). Specifically, the government generally requires temples to install both water scrubbers and electrostatic precipitators on their furnaces. The scrubbers inject water into the flue gas to cool it down, thus controlling the fly ash and big particles of dust, while the precipitator removes the smaller particles of dust from the gas.

According to Danny and Duncan, the Hong Kong Products Council has developed a furnace that contains scrubbers and precipitators, and it's available for about twenty thousand US dollars, an amount that Duncan told me the temples are willing to spend.

According to the guys, the agency works collaboratively with the temples to make sure the temples know what they have to do and how to do it, and the temples invariably respond positively both because of the possibility of legal sanctions and because of Hong Kong society norms, which push people toward compromise and conciliation. This last part was pretty interesting. My conversation with Dr. Tong from the Taoist Association suggested that the law, rather than any of these so-called norms, was what motivated the Taoist temples to install the required technology. Whatever is at work—probably some combination of law and norms—the system seems to be fairly successful. According to Danny and Duncan, the agency has never fined a temple under Section 10 of the statute.

It was great talking to the Taoists, the scientist, and the funny guys from the government, but by the end of all the interviews, I had had enough yapping and was looking forward to seeing some serious smoke. The next day, I got my wish. The morning started rough because the night before, I'd eaten a mussel that tasted like an armpit, but before long, I was back on the subway and off to see my first temple of the day. Wong Tai Sin is one of Hong Kong's biggest, busiest, and most famous temples, where thousands of worshippers come every day to pray, burn incense, and have their fortunes told. I came on a weekend afternoon, so the place was completely chaotic—packed with people, earsplitting, and smoky. Everywhere I looked, worshippers were chanting and bowing and kneeling and praying and, inevitably, clutching bouquets of incense sticks, whose plumes of smoke mingled together to create a toxic fog that was impossible to avoid. Having just heard about the health effects of incense and joss-stick burning the day before, I tried to stop breathing, but, alas, that turned out to be impossible.

Actually, there was one place within the complicated maze of buildings, staircases, and statues where it was possible to breathe completely smokeless air. Located off to the side, near the restrooms and the line of barbecue-grill-like things where people come to light their incense sticks, is a little ticket booth and a short set of stairs going down into what looks like the basement of the temple. For one hundred Hong Kong dollars (about twelve US bucks), I bought a ticket to Sik Sik Yuen, an underground circular chamber that looks like something out of a science fiction movie. Lit with indirect, space-age, glowing blue light, surrounded by seemingly hundreds of small statues of various deities, and featuring a large inlaid yin-yang symbol in the center of the floor, this hall was definitely the environmental jackpot that I had been looking for. It was also one of the strangest places I had visited in all my travels.

The temple had added the room a few years ago, at considerable cost, as a way of promoting environmentally friendly worship. As I found out, moreover, if you pay the one hundred Hong Kong dollars, you are pretty much *forced* to worship in an environmentally friendly way. When I arrived in the hall, a lady came up to me and opened the little pouch that I had received along with my ticket. In the pouch were three incense cones and a little amulet, which the lady arranged on my hand in some specific way that I did not understand before leading me to a central incense-holding place, where she told me to place one of the incense cones. I did what she said, and then she took me to one of the god statues, which she explained was the monitor god for the year, and I placed another incense cone in front of him. Finally, she asked me what year I was born, which then led us to a different god, where I placed the third incense cone. Nobody lit any of the incense, and when I had done all the things I was supposed to do, a guy gave me a tiny red pouch with a coin in it, which probably signified something, although I don't know what.

When this was over, the lady left me alone to wander around the room. The only other people present were three serious-looking guys wearing deep-red robes and tall black hats. After a while, one of them came up to me and started chatting. He told me his name was S. Y. Chu, and he wrote it in both English and Chinese in my notebook. Explaining that he was a Taoist priest, he asked me if I too was religious. I told him that I'm not religious now, but that I was raised Jewish. His eyes got really big when I mentioned this. He asked for my notebook again and wrote the word "Jevish" in it. I asked him what was going on with this part of the temple, and he said it was a "green temple." I mentioned that it seemed empty, while outside, the place was packed, and he said yes. I said, "It's kind of expensive to get in here," and again he agreed.

We stood there for a minute, S. Y. Chu and I, looking around at the place, and as I tried to think of something to say to fill the awkward silence, I thought back to the newspaper article I had read about the green temple. In the article, I recalled, there was a paragraph about how when a worshipper does something in the temple, something happens. I couldn't remember what either "something" was, but I had some inkling that when a worshipper puts the incense into the central holder, maybe a symbolic puff of smokeless smoke comes out of somewhere, or maybe a sound goes off to symbolize that the offering had been received. Nothing had happened like this when I put my little cone of incense anywhere, but this didn't stop me from saying to the Taoist priest, "So when you place the incense in the right place, there's like a *beep* sound, huh?"

Mr. Chu was astounded by this. He turned to look at me, his eyes growing even larger than when I had said I was Jewish, and asked if I had really heard a sound when I placed my incense. "Did you have religious experience?" he asked, hopefully.

I shook my head and apologized. "No," I said. "I'm afraid not."

As I emerged from the silent and pristine underground chamber back into the chaos and noise and smoky fog of the temple above, I felt the same way that I had felt at the little Singapore temple with the brand-new smokeless golden pagodas that nobody was using.

Sure, the thousands of people who were bowing, chanting, and lighting incense at the above-ground temple were making a big mess of the local airspace, but the place was vital in a way that the lovely cave was not. What good is it to have a place where people *can* worship without polluting the air if it's going to be so sterile (not to mention expensive) that nobody *wants* to worship there? I definitely admire what the temple did by building a place for worship where people can breathe without putting their lives at risk, and perhaps over time such temples will catch on. For now, though, I just found the setting odd and a little depressing.

After leaving Wong Tai Sin, I made my way to the Yuen Yuen Institute, which had been recommended to me by the Taoists as a spot where I could lay my eyes on some real, old-fashioned, unregulated joss burning. Located way out from the center of the city—a taxi ride from the last stop on the subway's Red Line—the institute is an enormous, sprawling place filled with temples and pavilions and prayer halls and other buildings and areas dedicated to all three of the country's major religions—Buddhism, Taoism, and Confucianism. Tong Wai Hop had told me that on the day I was going to visit, which was a weekend day during the Hungry Ghost festival, maybe a million people would be visiting the institute to pray and burn joss paper. And when my taxi pulled up to the entrance of the place on the top of a large hill, I could see that he was probably right. A lot of people had been at Wong Tai Sin, but this was another order of magnitude. And unlike Wong Tai Sin, there were no tourists to be seen. Among the thousands and thousands of people arriving at the institute, mine was the only white face around.

Because the institute is so large, I had no idea where to go or what I should be looking for. I wandered around the smoky grounds (like at Wong Tai Sin, it seemed as if everyone was carrying around some smoldering sticks of incense), popping into watch a Taoist prayer ceremony here, admiring a Buddhist statue there. I stood for a while at what seemed like the main center of the institute and watched the people. Almost everyone had arrived with at least one big bag of joss paper. Some families had several big bags among them. Somewhere in this crazy, smoky maze, there had to be a place where all the people were bringing their joss and burning it. I just couldn't figure out where it was.

So I started following people. I had a few false starts—one old guy heading to the bathroom, a family that put their bags down to light incense for some god or other—but before long, I had fallen behind a tiny old lady who was definitely dragging her two overstuffed bags of joss paper to be burned somewhere. I sauntered lazily behind her, watching her struggle mightily with the two unruly sacks. She turned a corner onto a wide road leading up a hill that was filled with similarly minded people carrying their bags of paper to the promised land. About a week after arriving in Asia, I knew I was finally hot on the trail of a truly huge burning extravaganza. Eventually I found myself in a clearing with hundreds of people and a monstrous, towering, filter-less black furnace, its three smokestacks barfing out smoke and ash into the atmosphere like some kind of angry dragon. *Jackpot!*

The place was astounding. I stood off to the side and, for as long as my lungs could stand it, watched as people old and young, male and female, approached the furnace, held their brightly colored sacks of joss in front of them, waved them up and down like the bags were bowing to the furnace, and then threw them in, where the stuff was promptly engulfed in hot flames and turned into dark plumes of carcinogenic smoke. I thought to myself, *Not a good day to wear these white linen pants.* Some people threw in only one package. Others tossed in several sacks. One group even brought a smartly constructed paper castle that took two men to hurl into the fire. I lifted my gaze to the top of the three smokestacks and watched the dense soot and scraps of half-burned paper fly into the sky and get carried away by the wind. I was glad that the wind that day was blowing away from the institute rather than toward it, or I might have had a cardiac infarction right on the spot.

From my travels in Singapore and Hong Kong, I learned a number of things about religious practices that harm the environment. First, the pollution from joss burning, which, like many other religious practices, harms the environment, justifies at least some government regulation. When faced with a substantial risk to the environment and human health, the government has an obligation to regulate activities to protect its citizens, regardless of the religious motivation of the activity. Observing and breathing in the smoke coming out of many of these furnaces, I felt certain that the practice of burning joss paper is dangerous, not only to those who personally engage in it but also to the bystanders who are basically forced to breathe in the dangerous, carcinogenic smoke sent up by all the burning. Joss paper is hardly the biggest contributor to air pollution in the region, but this detail is beside the point. In localized areas, and at certain times, the bursts of pollution from temples and other burning sites can be substantial, and when it comes to pulmonary issues, these kinds of short bursts can cause significant problems. In the United States, for example, the American Lung Association has long supported stricter air pollution measures to protect against bursts of pollution as short as five minutes long.

Second, when it comes to regulating religion, if it is possible, finding a technology-based approach will usually be preferable to outright bans. With joss burning, for example, the approach taken by the Hong Kong government—requiring temples to install scrubbers and other technology in their furnaces—is one that makes a lot of sense and that others should follow. Complete bans make it impossible for believers to practice their faith (or, worse, force believers to practice in secret, where regulation may not be able to reach them) and are therefore both unfair to believers and likely to backfire politically. The reaction of the community to the local government's attempt to ban joss burning, and the Singaporean prime minister's need to intervene in this issue, is a perfect example of regulation's going too far. Technological approaches to decreasing pollution, though inevitably only partial solutions, are a hallmark of modern environmental regulation. In the United States, for instance, both the Clean Water Act and the Clean Air Act require factories and power plants to impose certain technologies to reduce the pollution they emit rather than prohibiting the discharge of any pollution as a way of balancing the costs of regulation to industry with the benefits of reducing pollution. Given the large stakes on the religion side of the environment-religion balance, technological solutions should play at least an equally big role when the government seeks to reduce pollution caused by religious practices.

Third, the government should recognize that regulating religious practice will have significant impacts on the religion itself and should seek to minimize these impacts. Of course, many aspects of modern culture and society, including law, will inevitably affect how individuals practice their religion, but the government should attempt to keep its own effects to a minimum if possible. Having watched Taoists and Buddhists in three countries in all sorts of places (near their homes, in temples large and small, on the sidewalks outside their stores) burn joss paper to please their ancestors, I have learned that this is a practice that many, many millions of people take very seriously. The practice is part of these believers' identities as individuals and members of their communities, and they believe that burning the paper is necessary to keep angry ghosts at bay and their ancestors happy. Those of us who are not religious, or who are religious but are not Taoists, may scoff at this idea, but spend one afternoon at an enormous monastery where a million people have come on a single day with great seriousness to burn paper in a giant furnace, and you will realize how big a deal this joss burning really is.

Even if the government thinks that the effects of its regulation will be minimal, it is likely that the regulation will in fact have substantial effects on how the religion is practiced. As Professor Tong in Singapore explained to me, the regulation of joss paper burning in his country has significantly changed the way that Taoists practice their faith, essentially turning a private family ritual that took place at home into a public one, where groups of people now come together in public spaces like street corners and temples and use shared bins and furnaces rather than private receptacles. The point is that we cannot ignore the costs that religion will incur from regulation. This isn't to say that regulation won't be justified—only that government officials should remember these costs and try to minimize them if they can. Regulating those who practice their religion is very different from regulating, for example, a business that is producing nickel washers for sale in the marketplace. Regulating the nickel producer may cause the business to earn less money, but it will not affect the way that the producer fundamentally relates to what he or she believes is most meaningful in life. Regulating how a believer practices his or her religion does exactly that.

Fourth, the government, when crafting a technology-based regulation, should make sure to avoid requirements that will thoroughly reduce the vitality and vibrancy of the religious practice. At the extreme, even technological solutions have the potential to alter the religion so much that it becomes nearly unrecognizable and ultimately empty of any life. During this trip to Asia, looking at the sad, smokeless joss burners in Singapore and the space-age, smoke-free worship chamber in Hong Kong, I first started thinking about the danger of, for lack of a better word, *neutering* religion. The examples I witnessed in East Asia (similar to the example of the artificial pool for immersing idols in Mumbai) were the result of the changes made by religious institutions themselves, but one can easily imagine a government regulation that has similar results. Does anyone really want to turn religious practice into a quiet, pallid, lifeless affair? Such a result would only be justified if it were absolutely necessary to protect against particularly egregious environmental results. In my opinion, the air pollution caused by joss burning doesn't reach that level. A little smoke is bad, but it's not *that* bad.

Finally, to reduce negative effects on religion while still decreasing environmental harm, the government should work together with religious individuals and groups while drafting and implementing *its* regulations. The importance of such a relationship would seem to be one of the lessons that we can learn from Danny

Lo and Duncan Wong, who ascribed some of Hong Kong's success in getting temples to implement technological changes to the collaborative nature of the regulatory process. Such collaboration is in fact fairly common in regulatory environments outside the religion context, and there's no reason to think that religion should be treated any differently. In other words, just because the secular government is planning to impose requirements on a religious community does not mean that the government should keep the religious community at arm's length. Learning the needs and fears of the community should help the government craft a regulation that will work to protect the environment without unduly alienating or harming the community. Of course, collaboration may not work as well in some cultures as it does in others, but even in a litigious culture like the one we have in the United States, there is reason to think collaboration will help. As we have seen in the example of the Pennsylvanian Amish, for example, the government made far more headway when it started working with the religious community than when it came in "with guns ablazing." When it comes to regulating religious practices to protect the environment, the government should definitely put its guns away.

TAIWANESE TURTLES

*Religious Authenticity
and the Role of Government*

When the famous monk announced that it was time to go outside, my heart leaped. I had spent the past eight hours inside a hilltop temple in the small town of Keelung just north of Taipei, trying to chant Buddhist scripture and performing a variety of rituals that I hadn't understood, so I was thrilled by the chance to stretch my legs and get some fresh air. Plus, if the inside part of the temple ceremony was over, that meant it was getting closer to the time when the monk would lead the two hundred or so parishioners in the practice of "mercy release," which is what, after all, I had come to the temple to see.

I stood up and followed the directions of the super nice middle-aged Chinese lady named Jen, who had been helping me figure out what was going on all day long. Unfortunately, what I had to do next was no easy matter. Jen told me I had to bring with me not only the two pieces of white bread that I had squished around strands of my own hair, but also the little cup of salt water, the packet of jasmine, and the prayer book. These were hard enough to juggle even before Jen informed me that I could not let the impure bread touch the book. Or was it the book that was impure? And could the

bread touch the jasmine? I wasn't sure. Just to be safe, I tried holding all the things without any of the things touching any other of the
things, but then someone handed me two burning sticks of incense,
and any chance of holding all the things the right way went out the
window. And then it started to rain. When my too-big khaki pants
started to fall down, I felt like throwing everything on the ground
and sprinting for the exit.

Raise your hand if you've heard of mercy release. That's okay—I
hadn't heard of it either until I started researching this book, but it
turns out to be one of the most bizarre religious practices I've ever
learned about. The practice has its roots in ancient Buddhist scripture, which teaches that if you see an animal that has been captured
or is trapped in some way, you should release the animal, both for
the animal's own sake and to improve your own karma. Some Buddhists still practice mercy release in this way, but these days, some
Buddhist temples have turned this simple, admirable activity into
big business. They hand over big bucks to buy huge numbers—we're
talking thousands, hundreds of thousands, millions—of small animals, mostly fish, birds, frogs, and turtles, which the Buddhists then
release into the environment, where, if the creatures haven't died
already, they are promptly captured again and sold back to the temples and groups to be released once more. In return, the religious
organizations receive tons of money in donations from parishioners
for the opportunity to participate in the ritual, which many believe
will bring them good karma in their future lives.

Although releasing a couple of animals into the environment
here and there won't likely harm the environment at all, the veritable "karma mill" that has been created by these large Buddhist
organizations does cause significant environmental injury in a couple of ways. First, many of the animals die while being captured,
stored, transported, or released. According to one estimate, over
two hundred million animals are released every year in Taiwan

alone, and many of these are injured or killed in the process. The Humane Society International describes some of the harm that the animals can endure:

> Animals trapped for mercy release can sustain fatal injuries from the nets or snares. Others suffocate or starve during transport, when they're kept in tightly packed crates for days or weeks. Animals who survive to be released often collapse from exhaustion, illness, or injury or become easy prey for predators. Many die after they're released into inappropriate habitats. For example, freshwater turtles may be let loose in the ocean, and saltwater fish may be placed in ponds or rivers. In some cases, hunters wait just beyond the ceremony sites to recapture the animals so they can be resold.

The "karma mills" can cause significant environmental injury in another way. Because many of these animals are released into unsuitable habitats, they can harm those habitats or even destroy them, along with the other animals that call these habitats home. Again, Humane Society International describes the problem: "Mercy release can also harm ecosystems. Animals may be released outside their natural ranges, sometimes thousands of miles from where they were captured and in groups large enough to establish breeding populations. They can spread diseases to native species, compete for food and territory, or mate with indigenous animals, threatening gene pools." One of the rare scientific studies on the effects of mercy release, for example, studied the effect of the practice on American bullfrog populations in Yunnan Province, China, and concluded that "religious release is an important pathway for wildlife invasions and has implications for prevention and management on a global scale."

To learn what I could about mercy release, I had chosen to visit Taiwan instead of the many other countries where it takes place—Thailand, mainland China, Australia, even the United States—because Taiwan is where the practice has become most prominent and controversial. I had also lived in Taipei for a year after college, and I was curious to see how the place had changed in the past twenty years. My memory was that it was a chaotic and polluted city that was an incredibly fun place to live as a young person who liked staying up late, dancing to 1990s grunge, and drinking cheap booze. But who knew what I would think of it as an old person who likes getting up early, listening to jazz, and drinking cheap booze? In addition to getting a better understanding of mercy release, I also had set myself a somewhat difficult goal to achieve in the six or so days that I was going to spend in the island nation, namely, to witness firsthand someone or some group of people actually engaging in the practice. I had no idea how I was going to go about attaining this goal, short of sitting by some random river and waiting to see if anyone would come by and throw a thousand turtles into it. I did, however, have a couple of contacts who could probably point me in the right direction, so I showed up in Taipei and hoped for the best.

I first learned about mercy release while communicating over e-mail with a Taiwan-based writer named Steven Crook. Crook explained the practice to me, and although I could hardly believe it, a quick web search confirmed its reality. It didn't take much more searching on the Internet to learn that the person to talk to in the United States about mercy release was Iris Ho, wildlife program manager at the Humane Society International in Washington, DC. For years, Ho has worked on all sorts of important and high-profile issues involving animal rights, such as protecting elephants from ivory hunters, stopping poachers who kill rhinoceroses for their horns, and convincing Chinese restaurants to stop serving shark fin soup. She is also the society's point person on the mercy-release issue.

When I talked with Ho on the phone in early 2013, she explained to me the vast scope of the problem. Apparently, practitioners release hundreds of millions of animals every year, resulting in countless deaths and enormous damage to aquatic and other habitats. She told me that although mercy release is widely practiced in mainland China, the practice is most openly studied and debated in Taiwan. Ho also said that public opinion in Taiwan is largely against mercy release, but that it persists nonetheless, with supporters claiming that the practice is not only within their religious freedom rights but also actually good for the country. Indeed, Ho told me that one prominent supporter had claimed that an enormous typhoon had recently missed hitting the island because mercy release was practiced there so often. According to Ho, although the practice is mainly carried out by big temples, lots of individuals do it on their own as well. On a recent trip to Taiwan, Ho said, she and some of her fellow animal rights advocates one day randomly ran into a bunch of people who had bought some fishes at a seafood restaurant and were releasing them into a nearby stream. Ho said that the releasers were "so happy" and tried to get Ho and her compatriots to join them. "Just do it," they said. "It's so good for you!" Ho and her compatriots demurred.

Most interestingly, Ho explained to me that to combat the corrupt and dangerous big-business model of mercy release, animal rights groups have started working together with more kind-minded Buddhist organizations to organize small-scale alternatives that truly embody the spirit behind the ancient mercy release teaching without causing any environmental damage. Recently, for example, the Venerable Benkong Shi, an American Buddhist monk from New York, and Lorri Cramer, a rehabilitation specialist with the New York Turtle and Tortoise Society, became aware around the same time that Buddhists around New York City had been releasing turtles in large numbers into inappropriate habitats, including Central Park. The two came together to plan events they call Compassionate Release, where they release turtles that have been rehabilitated from injuries back into the wild. The ceremonies involve far fewer turtles and don't harm the environment. Cramer and Shi hope that Buddhists who practice the dangerous version of mercy release will begin participating in their ceremonies instead. According to Ho, the main theme of a recent major conference on the issue of mercy release that was held in Taiwan centered on developing alternatives like this one, and apparently, the topic was going to come up again in a couple of months at a conference in Baltimore. I went right out and bought my ticket to the Charm City.

The conference was the twenty-sixth International Congress for Conservation Biology, and in case you were wondering, its mascot was a stuffed blue crab named "Clawdia." I went all the way to Baltimore just to attend a single session, which was loquaciously titled "The Impact of Animal Release on Biodiversity and Human Health: Exploring Opportunities to Bridge Conservation and Religion." I sat in one of the uncomfortable chairs and waited to see how many people would in fact show up for a panel on this obscure topic. As it turned out, attendance wasn't so bad—maybe twenty-five or thirty people came to hear the panel of seven discuss the problem of mercy release and what scientists could do about it.

Chairing the panel was Stephen Awoyemi, a Nigerian scientist who, among other things, was the chairperson of the Religion and Conservation Research Collaborative, a committee of the Religion and Conservation Biology Working Group of the Society for Conservation Biology. Awoyemi's collaborative had recently issued a policy paper and a related letter in the prestigious journal *Science*, urging Buddhists to embrace alternatives to the dangerous large-scale practice of mercy release, such as adopting a cow that was otherwise destined for slaughter. The publications also made it clear to critics of Buddhism that some Buddhists groups have already begun to take such steps.

But now, at the conference, the panelists gave a series of presentations uneven in their ability to hold the audience's interest. An academic argued that the current practice of mercy release is a deviation from true Buddhist teachings. A prominent wildlife advocate from the Chinese mainland gave a presentation on the grim situation there. The presentation included a picture of a temple in Shanxi, where bags holding over ten thousand snakes, one-third of which had already died in the bags, were waiting to be opened as part of a mercy-release ceremony. Two women from Taiwan—a Buddhist named Li-Yi from a group called Bliss and Wisdom International and a scientist named Fang-Tse (Elaine) Chan from the Endemic Species Research Institute—explained how their two organizations had worked together to organize over thirty small-scale releases of rehabilitated animals as an alternative to the awful large-scale releases that are prevalent all over Taiwan.

The highlight of the program, though, was the Venerable Benkong Shi, the New York monk. Shi is a big white guy in his sixties who looks, with his bald head and Buddhist garb, sort of like Jeffrey Tambor in a bathrobe. He speaks with a strong New York accent, and he definitely has a sense of humor about himself. When he learned about the turtles being released in Central Park and openly criticized the releasers as "ignorant," for example, he soon realized that "being a white man dressed as Kung Fu didn't help" him stop the practice. Shi explained that shortly after he had publicly criticized the Central Park release, he woke up one morning to find that someone had left him a tiny turtle at his door. Not knowing what to do, he called the Turtle and Tortoise Society and asked for advice. The staff told him to take care of it for the next eighty or eighty-five years, because that's how long turtles can live.

Shi was attending the conference as a kind of ambassador to the scientific community from the religious world. Stressing that mercy release has gone beyond Buddhism and entered Chinese culture more broadly (indeed, one academic paper I read says that even some Christian churches in Taiwan have practiced it), Shi explained that the believers who do the mass releases truly, deeply believe that they are helping and that good things will come back to them from releasing the animals. But he said that temples are happy to learn how to work with conservationists and that conservationists should reach out to them. "Let's work together," he said. "Knock on our door. We won't shoot you."

I was lucky enough to be invited to join Iris Ho and the panelists at lunch afterward and even luckier to be seated near Elaine Chan, who was extremely friendly and funny and happy to tell me what she knew about the mercy-release situation in Taiwan. While negotiating the grotesquely US-sized portion of food on her plate, she told me about the king of all mercy releasers—a Taipei-based monk named Hai Tao. The monk's enormous following owes itself not only to his personal charisma, but also to the fact that he runs a twenty-four-hour-per-day television station called Life TV, which allows him to easily spread his message around the world. According to Life TV's website, "Besides broadcasting the orthodox Buddhist programs and advocating the values of no killing, life release, Buddha-remembrance and vegetarianism, Life TV also serves as the great information integration platform for all Ven. Hai Tao's missions. . . . Through 24-hour broadcast, Life TV intends to allow all of the sentient beings, visible or invisible, to benefit from the Dharma teaching." It was clear from the way that Elaine (and others at the table) talked about Hai Tao that a lot of people think he is ridiculous. Not to mention dangerous. Kind of like how we liberals in the United States used to talk about Jerry Falwell. Hai Tao, in other words, is the mercy-release reformers' public enemy number one.

Elaine also told me that if I made it to Taiwan, I could come to the town where she works—maybe a hundred miles south of Taipei—and she and other people from her office could drive me around to try to show me some mercy release in action. Perhaps, she further suggested, her office might be doing one of their rehabilitation releases along with Bliss and Wisdom International while I was there, and I could check that out. As it happened, a few weeks before I made my trip, Elaine told me by e-mail that indeed they would be releasing a rehabilitated macaque while I was there. Even though I wasn't sure what a macaque was, I was psyched that I would actually get to see a good-style animal release in Taiwan. Imagine how much more excited I was when I learned that a macaque is not a small, colorful bird, as I thought, but a small, funny-looking monkey!

I arrived for my week in Taipei after my visits to Singapore and Hong Kong, and I quickly realized that the city was wildly different from how I remembered it in the early 1990s. For one thing, there is now a subway, which has completely changed the place. It used to be that getting around town was nearly impossible; the traffic there would make someone yearn for the calm, orderly streets of Mumbai. Now the city is both much easier to navigate and a whole lot less polluted. Sadly, most of the bars and clubs that I remember from my time living there were gone, although it is doubtful I would have ventured onto the dance floor anyway, because I'm not sure that people over the age of thirty-five are actually allowed to dance anywhere these days other than at weddings. Although I didn't recognize either the traffic or many of the bars anymore, the August weather was familiar—the temperature regularly hovered above ninety degrees, the humidity was approximately 173 percent, and a fierce typhoon was bearing down on the island.

My first day in Taiwan was devoted to meeting people and talking about mercy release. First up was Wu Hung, the world-famous animal rights advocate who runs the Environment and Animal Society, or EAST, the leading anti-mercy-release organization in Taiwan. Wu, or Chu Tseng-hung, as he's also known (Wu Hung is his Buddhist name), had been a Buddhist monk before deciding to devote himself full-time to the protection of animals. In his role as the head of EAST, Wu has worked on issues like protecting domestic animals from abuse, encouraging the humane slaughter of food animals, cracking down on the sale of bear bile, and promoting the safe operation of Taiwanese zoos. (In 2007, a crocodile at a zoo in Taiwan's second-largest city bit the arm off of a veterinarian.) When I visited him, he and his coworkers were preparing for a press conference the following day. Wu intended to criticize the government's plan to infect a bunch of beagles with rabies so it could study how the disease spreads among different species. The government had hatched the plan because of the recent reemergence of the disease in the nation, and the proposal had quickly attracted much negative attention from activists around the world, including, of course, Alec Baldwin.

When I arrived at Wu's office, exhausted and dripping with sweat from searching the nearby streets for the right address, I was promptly greeted by a white and black office cat named Jing Jing and provided with a nice tall glass of water. Wu came in shortly after, sat down at the table where I was also sitting, and promptly smashed a small bug into oblivion on the table's surface, which I thought was odd.

Since Wu had been a monk, I was curious whether he thought there were any good arguments from within the Buddhist tradition in favor of the big-business model of mercy release. According to Wu, who many years ago as a young monk had even engaged in the practice, releasing frogs into a river, the practitioners of mass mercy release like to quote the sutras to defend their actions, but they usually either change key words in the passages or otherwise explain the sutras incorrectly. He said that although many Buddhists continue to participate in mercy release on a large scale, "most disagree that this is how you should practice your beliefs."

Wu explained to me how the government in Taiwan—at both the local and the national levels—has started taking steps to regulate mercy release in the country. A couple of large cities apparently have clear and specific regulations for releases. At least one other midsized county has adopted a purely symbolic "regulation" that has nonetheless helped decrease the practice within its borders. A few years ago, the national government gave a large grant to Li Mau Sheng, a law professor at National Taiwan University, to prepare a report on the nature and extent of the practice within Taiwan. On the basis of Dr. Li's report and investigations carried out by EAST, the Forestry Bureau has issued a proposed regulation that would give the bureau the power to closely regulate mercy release throughout the entire country.

Still, Wu recognizes that the big money involved in mercy release will make it hard to crack down on Hai Tao and others like him.

With events like the Ganesh festival in my mind, where it seemed that the law could only play a minor role in reducing the environmental impact, I asked Wu if this was a problem that required legal intervention. He was quite clear that the law was needed. "This is a business, not a Buddhist practice," he said. "And if it's a business, then we need a law, we need to regulate it." Moreover, with the amount of money at Hai Tao's disposal, Wu worries that even the legal system might not succeed. "He's a businessman," Wu told me. "He knows there's a market; he will reject the legislation. The people with the money have too big an influence. It's a loophole in democracy." I completely understood what he was saying and indeed just barely stopped myself from flying into a rage about the Supreme Court's decision in *Citizens United*.

When I asked Wu if he thought I might have a chance to witness a mercy-release event during my week in Taiwan, he seemed doubtful. He asked one of his assistants to check to see if anything was planned, but the assistant initially came up empty. Wu said that usually, the big releases take place on the birthday of certain Buddhas, and there wasn't going to be any such birthday while I was there. Unfortunately, it seemed as though I had just missed a large and therefore incredibly dangerous release on the country's western coast—a release that involved over three hundred thousand fish.

But just as I was softly whining about my bad luck, Wu's assistant came up with something. It turned out that Hai Tao was in fact in Taiwan, just back from an overseas jaunt, and was going to be involved in some sort of ceremony the very next day. All that the office could come up with was a small advertisement from a website that featured Hai Tao's image and a few bits of information—the time and address of the event: eight o'clock in the morning at a temple in the district of Qidu in the city of Keelung. Wu said he had no idea at all what might be happening there, but maybe I should check it out. Who knows, maybe they would be releasing some unlucky animals, and I could learn something.

There was no way I was going to give up the possibility of seeing this infamous monk in person, so I decided that I would go and see what I could see. Plus, since I was also interested in studying joss paper burning in Taiwan, and since Keelung was supposed to be a place where a lot of joss was burnt, I figured I could kill two birds with one stone. Well, that might not be the best way to put it, but you get the idea.

Later in the afternoon, I took a cab to the campus of National Taiwan University, or Tai Da, the top university in the nation, to talk with Dr. Li Mau Sheng, the law professor who had written a massive report on the practice of mercy release for the government. It didn't take long to realize that this guy was a bit quirky. If the sign that read I LOVE SMOKING on his door, for example, hadn't tipped me off, it would have soon dawned on me when the professor started popping out of his chair and practically dancing with excitement while talking about his report (apparently, there are actually *two* places where old people can dance—weddings and in their own offices). Funny, garrulous, and obviously quite brilliant, the Danny-DeVito-ish professor explained to me the nature of his work and his predictions for whether the legal system would be able to control the abuses of mercy release.

Unfortunately, I didn't really understand much of what he said. My Chinese having gone to hell in the past twenty years, Mr. Wu had kindly agreed to translate for me, but even though he did a great job, a lot was lost in translation, and my notes are a bit of a scramble. A couple of main points, though, were clear. First, part of the reason Li worked on the report (other than being asked to by the government, which enjoys a close relationship with Tai Da) was that he feels empathy with the animals that, as he put it, "are released and then die." Second, Li supports a focused effort to deal with the problem—a single law administered by a single agency, rather than a networked or piecemeal approach in which different agencies attack different parts of the problem (transportation, care and handling of the animals, etc.) under different legal instruments. The government apparently agrees with him on this point, even though, as Li suggested to me, the worst abuses of mercy release might very well be controlled—theoretically, at least—through the application of various existing laws administered by different agencies.

Finally, however, Li is cynical about whether having a law controlling mercy release on the books will have any practical effect. According to him, the large government bureaucracy in Taiwan, which is responsible for enforcing the laws passed by the legislature, is insufficiently educated, understaffed, and largely ineffectual. Li makes an important point about the power of law to effect change—the law on the books is one thing, but the law in action is something quite different. Still, Li suggested that perhaps just having the law on the books would do some good. As he put it, "The government is preparing a weapon to scare Hai Tao. Maybe he will reduce his practices or move them overseas."

For my last meeting of the day, I headed over to the Bureau of Forestry to talk with some of the government officials who were championing the national law that is based on Professor Li's work. As with almost all my encounters with government officials during my travels (except for the unnerving Mumbai cop), I found the four people I met with at the bureau to be friendly, helpful, and generally devoted to doing the right thing. I sat down in a pretty typical government office to sip tea and talk with Lin Kuo Chang, the chief of the Wildlife Conservation Section, and three other bureau employees to find out what they thought about the law's necessity, its prospects for passage, and the likelihood of its success. Perhaps unsurprisingly, they seemed far more optimistic than Professor Li had been earlier in the day.

As for necessity, Lin and his fellow foresters agreed that the country needs a law. According to them, the number of animals being released has increased dramatically over the past ten years because of the commercialization of the practice. Lots of people, they emphasized, are making way too much money in the mercy-release business by capturing, transporting, and selling the animals, and of course, a temple collects even more money from its parishioners when it stages a mass release. When I asked the group what the Buddhist organizations say about their releases, a super sweet fifty-something lady with a blue flowered dress laughed and answered, "They think that the more the better—they think they are doing a very good thing for Taiwan." In response to my question about whether anyone has raised a religious-freedom objection to the proposed legislation, Mr. Lin told me that no one had raised it as an issue yet, but added, "It'll come up when we fine them." The proposed fines are indeed steep—up to fifty thousand new Taiwanese dollars (roughly fifteen thousand US dollars) for an unauthorized release, and even more if the release damages the environment. No wonder many religious groups are opposed to the law and insist that they should be able to self-regulate without government interference as they have done historically. "I don't want to say they're stubborn," said one of the women at the table. "But they don't want to accept other people's views."

The government officials recognize that they're dealing with a sensitive issue. "We're trying to balance religion and the environment," said the same woman, "trying to find a way in the middle. We're not against religious culture." They are optimistic, though, that they will succeed in getting the law passed. Not only is the public on their side, but many Buddhist groups are as well. The government hopes to cooperate with Buddhist groups and encourage them to do more of the small-scale eco-friendly alternative releases. Moreover, like everyone I talked with that day, the four government workers thought that just the process of introducing and debating the law would do some good by increasing awareness of the dangers of mercy release and getting the issue out into the open.

If mercy release had really been responsible for an earlier typhoon's missing the island, as that one Buddhist had told Iris Ho on her visit to Taiwan the year before, why then, with mercy release becoming more and more prevalent in the past couple of years, was there a giant typhoon making its way right for Taiwan when I was there? The typhoon was predicted to hit in two days, which, unfortunately, was the same day that I had planned to go south to see Elaine Chan and the nice Buddhists release the funny monkey. When I got home from my long day of interviews, I was therefore not surprised to find an e-mail in my inbox from Elaine telling me that the release had been canceled. The change of plans made me sad. Admittedly, however, it probably made the monkey even sadder.

I still had one day until the typhoon arrived, so the next morning, I made my way to the smelly main Taipei train station and headed off to Keelung to try to get a look at the notorious Hai Tao. As I walked the mile or so from the Keelung station through the light drizzle to the hilltop temple, I wondered what I might find when I arrived. I seriously had no idea whatsoever of what was awaiting me. The temple wasn't on the radar of any guidebook or English-language website, and the little advertisement that Wu Hung had given me about the event offered no clue. I tried running the Chinese from the ad through Google Translate and got the following, which was not helpful:

> Haitao Master Chi Yan Temple in Keelung France 8/20 8:30 organize Eight Precepts, nursing students, lighting, medicine for cum compassion soup puja; at 16:00 on the 21st successful mid-year Jizo France cum Kai Kenda Monsanto soup puja; respectfully Haitao master Principal Act, address: Keelung District, No. 48, Lane 65, new Street), Tel: 02-24,562,057.

I envisioned coming upon an intimate gathering of believers crowded around the charismatic leader's feet—some of the followers perhaps nursing students slurping soup—but in any event, a gathering where I would, no doubt, not be welcome. Approaching the temple hall where the event was taking place, however, I quickly realized that "intimate" would not be an accurate way to describe the gathering. About two hundred people were milling about the temple's gilded main room, most of them trying to find a good seat among the rows and rows of little blue plastic stools set out for that purpose. A surprisingly large number of workers wearing blue Life TV vests were assisting the parishioners, making last-minute adjustments to the many television cameras that would presumably be broadcasting the event, and otherwise making sure the proceedings would begin smoothly. A bunch of monks were seated in the front of the room. At first they were wearing brown robes, but at some point, they suddenly changed into black robes. The room was filled with flowers.

My hope was that I could hang back and watch inconspicuously from the sidelines or something, maybe leave if the ceremony turned out to be uninteresting or otherwise unlikely to involve the release of small animals into inappropriate environments. But as soon as I showed up, some older ladies who seemed happy to see me ushered me to a blue stool on the middle aisle about ten rows back from the front. I had little choice but to smile and thank them and take my seat. Soon the ceremony started with some group chanting and the banging of gongs, and I knew I was in it for the long haul.

Now, I've been to Buddhist ceremonies before. I've visited all sorts of Buddhist temples, from Bangkok to Hanoi to Tokyo. My wife and I even once stayed for several days in a Buddhist monastery somewhere in central Japan, getting up at five thirty in the morning to sit cross-legged during a lovely and peaceful ceremony led by the monastery's monks. This thing that I had gotten myself involved with in Keelung, however, was like nothing I had ever seen before.

For one thing, during the ubiquitous chanting, everyone was doing these choreographed finger movements that were impossible to follow. It was like being at a line dance but not knowing when to twirl. The biggest difference, by far, however, was all the props. I hadn't been sitting there more than ten minutes when someone came around to pass out tiny shot glasses filled with what looked like water and maybe a little salt. I had no idea what to do with my liquid, but what I was most worried about was that someone would make me drink it. Was this the proverbial Kool-Aid I'd heard so much about? I had a rough plan to splash the beverage on my forehead and confess my "drinking problem," *Airplane!* style, but it never came to that, because soon enough, after people had the chance to inexplicably put their fingers in the water and sprinkle it up into the air and on their neighbors, someone came by to collect the water glasses and replace them with hot little tea-votive-like candles. These were even more perplexing, because apparently there was a specific right way to hold them, and, of course, they were hot. I watched as everyone moved his or her candle around in a prescribed pattern while the chanting continued, and I tried to follow along, but I was hopeless. Soon, someone collected the candles and replaced them with packets of jasmine, which we rubbed all over ourselves. Some people were waving flags. Others had little plastic squirt bottles like the one I spray our cat with when she scratches the couch and were spritzing water into the air so the droplets would fall on everyone's heads. This went on for at least an hour before there was a break. I was confused and a little damp. Meanwhile, there were no little animals anywhere, and Hai Tao was nowhere to be seen.

During the break, a super-sweet middle-aged lady with a blue dress and long hair came and sat next to me. She spoke excellent English. She introduced herself as Jen, shook my hand, and promptly told me that we could not shake hands ever again because I was a man and she was a woman. For the next six hours, Jen would be my guide to this strange ritual. She was fascinated by the fact that a foreigner had shown up to see Hai Tao.

"Why did you come here?" she asked me.

I told her that I'd heard that the famous monk Hai Tao would be here and that I wanted to see him.

She looked at me weird. "You must have been a Buddhist in a previous life," she said. "You must have known Hai Tao in a previous life."

I nodded. "Maybe I did," I answered, since, who knows, maybe I did.

Jen explained that she came to see Hai Tao as often as she could, and when I faux-innocently brought up mercy release ("I hear that there's also a part of the ritual where you release lots of captured animals?"), she said that she often participated in such activities. But when I tried, ever so slightly, to press the mercy-release issue, she deflected the question and returned to how remarkable it was that I had shown up at this ritual at all. "This is precious time," she said to me. "Precious time."

For the next hour or two, the ceremony continued with a variety of activities—some sermonizing from the monk in the front of the room, a bit of call and response between the monk and the congregation, and some general chanting of sutras. I understood exactly zero percent of what was going on. At one point during the chanting, a chime rang from somewhere, and Jen instructed me to put my prayer book up in the air over my head and to keep it there until the chime rang again, at which point we all lowered our books to our laps. This happened maybe twenty times. When the morning activities were over, the group broke for lunch (surprisingly good and completely vegetarian), and after eating, I stepped outside to get some air and stretch a bit. Before long, a guy whose English name turned out to be York struck up a conversation with me, apparently intrigued by the presence of a non-Asian face at the gathering. Wearing dark glasses and a T-shirt devoted to an Indian guru, York also kind of stuck out in the crowd as being far younger and more casually dressed than most everyone else. He explained to me that he was a software programmer who liked to come to these types of ceremonies once or twice a month to "purify" his "heart and mind."

We talked for a bit about Buddhism, and I quickly moved the conversation to the practice of mercy release. He said he practices it regularly, with birds or fish or even worms. He explained that when he releases the animals, he always says certain mantras and parts of the sutras as part of the ritual and that he releases them with the hope that they will become Buddhas themselves. "Now, because they have done something bad in the past, they have been reincarnated low down," he said, "but I release them and hope that they will come back higher next time." It was fascinating to hear someone of my own generation—someone who was clearly highly educated—providing his perspective about the practice. Older people blindly following a charismatic leader aren't the only ones who engage in mercy release.

York said that he did not expect that there would be any mercy-release events on the day of the ceremony, and with that disappointing news, I was about to leave and head back to Taipei when Jen found me and convinced me to stay. Talking about Hai Tao, she was like a kid opening presents at a birthday party after too much cake and ice cream. "Hai Tao is such a good, good man," she exclaimed, only minutes after she had told me that he refuses to talk to women. She insisted that there would be a release after the ceremony—"there's always one"—and so I decided to stick it out a little longer.

After lunch, Hai Tao finally showed up. Bald, dressed in flowing golden robes, and wearing rimless glasses, he entered from the back and walked down the middle aisle toward the front of the hall, where he was met with the mass turning of heads and hushed gasps. The effect was sort of what I'd imagine it would be like if William Shatner suddenly arrived at a Star Trek convention. I watched eagerly with the rest of the group as the famous monk took his place at the dais at the front of the hall. Hai Tao ran the second half of the ceremony with a calm and mostly affectless demeanor that made it a bit difficult to understand why so many people are drawn to him. Jen told me that the Buddha had brought her to Hai Tao, which I wondered about, but I didn't really know how to frame a follow-up question that would further the conversation in a fruitful manner.

In convincing me to stay for the second half, Jen had told me that the afternoon part of the ceremony would be more "mixed up" and would involve more "moving around" than the morning session. But I kid you not when I say that with the exception of two short breaks, the congregation did nothing but sit and chant from a sutra book for the next two hours. For those keeping score at home, that's *134 freaking pages of sutras*. I sat on my blue stool in a daze, listening to the mesmerizing rhythmic sound of two hundred people chanting in unison in Chinese as I hummed along with the beat.

During one of the breaks, Jen approached me and said that if I did what the rest of the group did during the rest of the ceremony I would "get many merits." "In the future," she told me, "you will think of today—what we did today was get rid of all the bad stuff and bring in the good stuff. You're really lucky to be here. Wonderful!" She was so excited, and her enthusiasm was so contagious, that I found myself thinking that perhaps someday I *would* indeed be reincarnated as a Buddha, or as a bodhisattva, or at least as a successful hedge-fund manager with a huge boat.

The chanting finally over, Hai Tao took the congregation back through everything we had done earlier in the day, complete with the finger movements, the salty water, the hot little candle, and the jasmine packet. This time, Jen tried to help me do everything in the right way, and I probably succeeded about 22 percent of the time. The one addition to what we'd done in the morning involved the bread. Oh yes, the bread! Someone came around and handed each person two slices of white bread. Confused, since I'd been told that there would be no more eating, I looked around for some cheese. But no. This bread was not for eating. Jen directed me to pluck out a strand of my hair, put it on one of the slices of bread, and squish the two pieces of bread together. I followed her instructions and ended up with a mushy mess. Then Jen told me I had to mold the mushy bread mess into a little person. I looked at her incredulously. *Are you kidding me?* Jen was definitely not kidding. I did the best I could to make the bread look like a guy. The bread looked nothing like a guy. I put my head in my hands and wondered how my life had reached such a point.

And then, just as I thought I had hit the zenith of weirdness, people started rubbing their little bread guys all over themselves.

Jen said that "we touch every part of our body" with the bread, and I think she said that it had something to do with promoting our health. All right. If you say so. I patted myself with the bread on my arms and my legs and my chest and then was going for my head when Jen stopped me. "We don't touch the head! Never the head!" *Holy crap! Imagine if I had touched the head?!* I'd almost blown it. I lowered my bread and reapplied it to my shoulders instead.

After the chanting, Hai Tao gave what seemed to be a sermon, and then he answered a series of written questions from the group. Perhaps he's allowed to read questions from women even if he can't talk to them? After this, Hai Tao announced that it was time to go outside. I was so excited that I could finally leave the little blue stool behind and perhaps go see some animal releases. First, though, more rituals! We stopped as a group outside the temple and turned back to the Buddha statue inside the temple to worship it with bowing and chanting and incense offering. As we moved from there to another location on the temple grounds where there was a large table covered with edible offerings for the Buddhas, workers came around and took one of our incense sticks, the saltwater dish, and one of the pieces of bread, leaving us with the book, one incense stick, one piece of mushed-up bread, and the jasmine packet. That was still a lot of stuff to be holding, but at least I was able, albeit with some difficulty, to pull up my pants. My victory, however, was short-lived, because as Hai Tao led the group in chanting over the food, some guy came over and gave me a flag to wave. I did my best to wave it while also holding all the other stuff. Soon enough, my pants started falling down again, and this time, I was helpless to stop their descent.

And that was it. Suddenly, the whole ceremony was over. No mercy releases whatsoever. No birds, no fish, no salamanders. Not even some worms or an amoeba. *What the hell?* I asked Jen what was going on, but all she really wanted to talk about was how much she wanted me to go and talk to Hai Tao, to meet him in person. She said that she would bring me over to meet him, but that she wouldn't be able to look him in the eye and would have to run away immediately. I demurred and asked again about mercy release. Finally, she relented. She said that, two weeks ago, they had let out ten gigantic truckfuls of fish into the sea by the northwestern city of Hsinchu.

I said: "That's a lot."

She said: "Yes."

I said: "Where do these fish come from?"

She said: "Breeders breed them for restaurants, but Master Hai Tao buys them to save them from the restaurants."

I refrained from telling her I thought that Hai Tao was pulling the wool over her eyes, and I simply said, "Oh."

I found York and asked him if he knew whether there might be any plans brewing for a nearby mercy release, and he pointed me to the bottom of the hill, where a midsized truck loaded up with plastic vats was idling. I immediately clambered down the hill and approached the truck. The vats were filled with fish, which I could hear splashing around inside. I asked a woman who seemed to be in charge of the truck whether the fish were going to be released today, and she said no, that they would be released sometime in the future near Hsinchu.

"Not today?" I queried weakly.

"No, not today."

The lady gave me a phone number to call that she said would inform me of when the next release might be, but when I got back to my hotel a couple of hours later and called the number, I couldn't decipher any of the information on the prerecorded message. All in all, despite being able to talk to some people who actually practice mercy release with Hai Tao, it had not been an overly successful visit—I hadn't seen what I had hoped to see. But perhaps my failure was simply a matter of timing. Had I come a week earlier or perhaps a week later, I could have witnessed exactly the kind of practice that has so angered the Taiwanese government and environmental groups around the world. Crestfallen and exhausted, I drank a few bottles of Taiwan Beer, crawled into bed, and watched the second half of *The Hangover, Part II* until I fell asleep.

For the larger part of the next two days, the typhoon battered Taipei with torrential rains and extreme winds. I stared out my hotel window, chowing on noodles and sipping beer purchased from the miraculously still open 7–Eleven that was right next to the hotel. I read a lot and watched many movies starring Seth Rogen. When the typhoon finally cleared, I had one more spot to visit. In an area of town near the bustling Longshan Temple, near Snake Alley, the notorious part of Taipei where you can go watch a snake be stripped of its skin and then drink a cup of its blood, is a stretch of Heping East Road known as Bird Street. The area, which I learned about from Wu Hung, is where you can go to buy any kind of bird you want—everything from a sparrow to a parakeet to a couple of lovebirds to a toucan to an honest-to-goodness buzzard—either to keep as a pet or, if you're so inclined, to release for karmic rewards.

Stinky, chirpy, and slathered in Avian flu, Bird Street is not a place I would recommend visiting, even though that thing about how it is slathered in Avian flu was a lie, probably. The birds (along with an occasional group of guinea pigs or giant snails) are kept in terrible conditions, packed into small ugly cages stacked on top of each other from floor to ceiling. The noise and smell were truly overpowering, and the air felt thick with poop and feathers. One small cage held eight ducklings; another housed a full-grown rooster. But the really sad part was the stacks of long, shallow wooden crates that were packed with sparrows. You purchase these if you want to have a big mercy-release event; the stores even have signs up that announce FANG SHENG NIAO ("mercy-release birds"). I don't know exactly how many sparrows were crammed into each crate, but it was a lot—way more than ever should have been packed into a space that tight. I approached one crate and said hello to the tiny birds. Though I didn't mention it to them, I secretly hoped that they would get to experience at least a brief moment of freedom in the future—sometime between when they are released from their tiny

prison and when they are recaptured and put right back into another tiny prison. I'm not sure how anyone could visit this place
and conclude that these large-scale so-called mercy releases are
even the slightest bit merciful.

The typhoon, as it turned out, was not the only thing that ended up
delaying the release of the rehabilitated macaque. Right around the
same time as the typhoon, Chan and her coworkers found a new
wound on the monkey's face and had to put off its release for a couple of weeks until the sore healed. And then, while the macaque
was rehabbing, a masked palm civet took its place in the queue to
be released, so the poor macaque had to wait its turn. (Chan's group
has only one release-training enclosure, and it is too small to ac-
commodate more than one animal at a time to prepare the animal
for reentering the wilderness.) Eventually, however, in late October,
the macaque was released.

Chan sent me a video that her office made of the monkey's long
road back to health and eventual release. Only someone with a
rotten eggplant in his or her chest instead of a heart could watch
the ten-minute film without feeling at least a tiny bit better about
the human race. The video starts with the discovery of the monkey, caught and injured in an illegal hunting trap in the middle of
the forest, in November 2012. The monkey is unconscious, with a
broken skull and injuries all over its body. From there we see the
animal rehabilitators working with the macaque to get him back
into shape—they feed him with an eyedropper-like thing and try to
teach him how to stand on two legs. Before long, the fuzzy gray guy
is chomping on solid food and walking around a cage, although the
scene where he first tries to walk out of the cage and falls down was
enough for me to reach for a box of tissues. By the end of the video,
the macaque is happily running around his rehabilitation enclosure and eating normal monkey food and even playing with what I
think is a potential mate. Finally, we see the monkey out in the wild,

scampering up a giant tree, soon to disappear back into the forest to live out the rest of his life in a natural habitat. Now *this* was truly a mercy release I could get behind.

One common way that nonbelievers rationalize or even justify imposing burdens on religious practice is by telling themselves or others that the practice in question is not *really* necessary or not even something that true Christians or Buddhists or whatever actually have to perform. Perhaps you've found yourself thinking something like this as you've been reading this book. Do Hindus really have to make twenty-five-foot idols to Ganesh? I know lots of Hindus who don't build giant elephantine idols, so maybe this isn't really a central practice of Hinduism after all. Can't Taoists burn one piece of paper instead of a thousand pieces? Can't the Jews in Israel light a candle on Lag B'Omer instead of a bonfire? My Jewish friends in the United States don't burn bonfires—surely, bonfires aren't something that you have to do to be Jewish, right?

Perhaps in none of the practices described in this book is it more tempting to press for abolition than with mass mercy release. After all, this is a practice that only a small percentage of Buddhists partake in; that dissenting Buddhists say relies on a misreading of the relevant sutras; that some scholars of Buddhism say is a deviation from true Buddhist teaching; and that does, by actually harming animals rather than helping them, seem to run counter to the Buddhist ideals most of us are familiar with. So perhaps the government can simply say that regulating this kind of mercy release, even banning it altogether, is completely fine because the ban does not infringe on real Buddhist practice.

While I was investigating mercy release and learning about its dangers, and particularly when I was in Taiwan looking at the stacks of crates stuffed with unfortunate little birds, I found myself thinking along these lines more than once. The more I thought about it, though, the less I liked the argument. Ultimately, I've decided that it's just plain wrong. The government should not justify regulating a religious practice by claiming that the practice is inauthentic, unnecessary to believers, or otherwise not a true reflection of how real members of a tradition practice their faith.

I reached this conclusion for several reasons. First of all, who is to say that one way of practicing any particular religious faith is true and that other ways are not? Religious believers disagree about what their faith requires all the time—consider, just as examples, the difference between Reform, Conservative, and Orthodox Judaism; whether the Episcopalian Church should allow gay ministers; and whether, as some Buddhists believe but others do not, laypeople as well as monks can reach enlightenment. The division of religious faiths into schools, the creation of sects, and the presence of schisms within faith traditions are indelible features of religious communities. A lot of Buddhists may think that Hai Tao has misread the sutras, but I'm betting that Hai Tao and his followers, including my very smart new friend York, whom I met at the temple, have a different view. Is there any objective method for determining who is right and who is wrong?

Second, even if it were theoretically possible to say that some interpretations of a religious tradition are wrong, an outsider to the tradition should not be the one deciding the matter. An outsider, such as a scholar, may be able to make a sophisticated argument as to what a contested portion of a sacred document means, based perhaps on linguistic or historical analysis. The conclusion, however, should hardly be binding on those who count themselves as part of the tradition, even if the conclusion might be of great interest to some of those people. How many Jews, regardless of what particular stripe of Judaism they might belong to, would agree to allow a Catholic, a Zoroastrian, or a secular scholar decide what counts as true Judaism? Ask yourself how you would react if a nonbeliever told you what your tradition means or doesn't mean. I certainly wouldn't want a Christian to tell me what I should believe as an atheist. I think it's fine if Buddhists debate the meaning of mercy release and how it should or should not be practiced, but it makes me deeply queasy when scholars or other observers who are not Buddhists start telling Buddhists what Buddhists really do or do not believe.

Third, and perhaps most importantly for purposes of this book, the government—which can employ the coercive power of law to radically change or even destroy a religious tradition—is the worst possible institution to decide on the true meaning of a given religion. Allowing a group of legislators, or agency bureaucrats, or judges to decide what counts as a *real* expression of a religious tradition and what does not is an extremely dangerous road to travel. At least in the United States, it runs deeply counter to our tradition of separating church and state—a tradition intended not only to protect the state from religion but also to protect religion from the state. If the state can decide what counts as real religion and what doesn't, then what's to stop it from defining some practice or belief as not being religious and then stomping it out?

For what it's worth, the Supreme Court has recognized this point on several occasions. For instance, it has ruled that although courts may decide whether a religious freedom claimant is being sincere, they may not opine about the truth of the claimant's beliefs. Moreover, in a case involving a Jehovah's Witness who claimed that the state had violated his religious freedom by requiring him to work in the armaments section of its factory, the Court held for the plaintiff even though other Jehovah's Witnesses had accepted work in that part of the factory. Chief Justice Warren Burger wrote:

> Intrafaith differences of that kind are not uncommon among followers of a particular creed, and the judicial process is singularly ill-equipped to resolve such differences in relation to the Religion Clauses. . . . Particularly in this sensitive area, it is not within the judicial function and judicial competence to inquire whether the petitioner or his fellow worker more correctly perceived the commands of their common faith. Courts are not arbiters of scriptural interpretation.

Indeed, in the two-hundred-plus years of its existence, the Supreme Court has never even attempted to articulate a definition of the word *religion* in the First Amendment. The other branches would do well to follow the Court's lead.

Finally, determining that some practice does not *really* represent the beliefs of a religious faith is not necessary to justify regulation to protect the environment. As I've explained elsewhere, the government may impose regulations even if those regulations happen to burden a religious practice. The government should try to be respectful toward the religion and minimize the harms to the religion to the extent possible, but it most certainly can (and should) take actions to protect the environment from danger. In the case of the type of mercy release practiced by Hai Tao and others like him, the rationale for regulation should be to protect the animals and ecosystems and should have nothing to do with whether this practice is really Buddhist or not. The question is whether the harm to the environment caused by the practice is sufficiently great to justify regulation, given the costs that it will impose on those who believe that mass mercy release is an important part of their religious practice. Reasonable people may differ on the answer to this question, but in my view, given the enormous numbers of animals harmed by the practice and the serious problems caused within ecosystems by the release of nonnative species, the balance should fall in favor of regulation. The Taiwanese government, in other words, is on the right track.

7

BARROW (ALASKA) BOWHEADS

Legal Exemptions and the Power of Community

It was about ten thirty in the morning in Barrow, Alaska—the northernmost city in the United States, over three hundred miles north of the Arctic Circle—when families started arriving at the Nalukataq grounds to celebrate the successful spring whaling season. The sun was out, which was not surprising, since it hadn't gone down for a month, but this morning it was particularly bright, and although the mid-June day wasn't exactly warm, it was not uncomfortably cold either. The grounds had been set up, as always, in the area just south of the Arctic Ocean and immediately west of the Top of the World Hotel, where I had been staying for the past week. The area was a little bit north of the lagoons that separate the older part of town in the south from the newer part, which is sometimes called Browerville. Tarps had been stretched across a series of tall wooden poles to create a wind barrier around the semicircular area, and the flags of the two successful whaling crews that were hosting the event were flying high over the people below. Each crew that catches a whale during the spring season is responsible for distributing whale meat to the rest of the community during the Nalukataq festival. The year I was there, the crews in town had brought in seven whales. This was the first of three days that the Inupiat people of Barrow and a few other nearby settlements would come together to celebrate both the whale hunt and the whales themselves, as the people have done for over a thousand years.

By eleven o'clock, the celebrants were arriving in droves—in the midafternoon, there would be well over a thousand people, most of them natives—and invariably they were carrying with them large coolers and rubber tubs, as well as full place settings, complete with cups, plates, forks, semicircular knives called *ulus*, and even salt and pepper shakers and the occasional bottle of Tabasco sauce. I, however, had none of these things. It wasn't because I hadn't been warned. The whole week, whenever I told someone I had come all the way from Boston to Barrow to witness the whaling festival, I was inevitably told to bring, at the very least, a bowl, a plate, and some cutlery. Others advised me to bring Ziploc bags. What I had actually managed to obtain, however, was far more modest—a small Styrofoam cup lifted from the hotel coffee station as well as a clamshell takeout container and a plastic fork and knife that I cajoled from a waiter at Niġġivikput, the hotel restaurant. These were just going to have to do.

I wandered around the grounds, taking in the scene and trying to predict what was going to happen. As the families arrived, they looked for prime spots around the perimeter where they could lay down their blankets, set up portable chairs, and settle in for the afternoon. Although there were a lot of people, most of them clearly knew each other, which was probably to be expected in a tight-knit community where there really isn't anywhere else to go. Toward the open end of the semicircle, the two hosting crews had set up tents; inside the tents, members of the crews hustled around getting ready for the celebration. A couple of long wooden tables had also been set up nearby and were covered with boxes of food and large metal pots filled with caribou soup and other concoctions that would soon be served to the entire community.

I looked with wonder farther into the semicircle at the so-called blanket toss contraption that would serve as the main source of entertainment as evening approached. Pulled taut between several wooden poles was a trampoline-like "blanket" fashioned out of the skins of bearded seals that had previously covered the two small wooden boats (called umiaks) that the successful crews had used to hunt their whales. Although, sadly, I wasn't able to stay around long enough to see it, in the evening the adults take turns bouncing on the blanket while the rest of the group grabs and snaps the thing, sending the bouncers flying high, high into the air. Back in the old days, hunters would do something similar to try to spot game far out onto the horizon, but now the toss is just for fun and some good-natured competition to see who can bounce the highest and most times without losing their footing or breaking a leg. Finally, in the middle of the semicircle, just sitting there on the ground on a piece of plywood, was a large chunk of meat that had obviously been part of a whale, though I had no idea what part of the whale it was from or what it was doing there.

Although I had hoped to run into at least one person whom I'd met over the past week, nobody looked familiar, and so I just walked around by myself, looking for a place to sit that would have a good view and not be on top of somebody else's stuff. I took a seat on the ground for a while, but it was uncomfortable, so I soon stood back up. A very old lady in a bulky arctic parka approached me— this was probably her eightieth or so Nalukataq—and told me that I should have carried a chair with me from home. When I told her that my home was four thousand miles away, she looked at me with a confused expression. I explained what I was doing at the festival, and then I asked her about the big piece of meat in the middle of the grounds.

"That's whale for visitors like you," she said. "But you will need a big knife or ulu to cut it with if you want to eat it."

"No problem," I replied, pulling my little plastic knife out from my pocket and presenting it proudly to her. "I have this!"

If you've never experienced an eightysomething woman cackling at you, eyes full of pity for your sad, silly self, let me just say that it can make you feel pretty stupid.

For whatever reason—perhaps their size, their gracefulness, or their intelligence—whales have always been objects of awe and admiration for human beings. We celebrate them in our books and movies and works of art and collectively spend two billion dollars every year to watch them swim, jump, and twirl around majestically in their natural habitats. Humans have also, however, long hunted whales, not only for their abundant meat, but also for their blubber, which used to be sold for lamp oil; their baleen, which some time ago was used to make ladies' corsets; and even their bones and smelly ambergris. By the mid-twentieth century, our seemingly unquenchable thirst for whale products had caused a massive decline in the numbers of whales and led to the listing of several species as endangered. The International Whaling Commission, a body created in 1946 to provide for the conservation of whale stocks, ultimately issued a complete ban on commercial whaling. The ban took effect in 1986. Although several countries still defy the IWC and engage in the practice, either outright or under the cover of doing "scientific research"—most notably Norway, Iceland, and Japan—the moratorium has certainly had a positive effect on whale populations around the world.

Although it prohibits commercial whaling, the IWC does authorize a smattering of communities around the world to engage in "subsistence whaling." The communities are places where people have traditionally hunted whales for their own use and where the whale hunt represents a critical aspect of the community's history and culture. Subsistence whaling communities exist in Greenland, Russia, Japan's eastern coast, Saint Vincent and the Grenadines in the Caribbean, and the United States. In the United States, the IWC regularly grants the Alaska Eskimo Whaling Commission the right to take a healthy number of bowhead whales every year (the actual number of whales taken varies widely, but is usually between forty and seventy-five), and the Alaskan commission distributes the quota among the eleven traditional whaling communities in the state, including Barrow. In recent years, the IWC has also allowed the Makah Tribe of western Washington State to hunt a small number of gray whales. Thus far, the Makahs have only taken one whale under the permit. As Robert Sullivan describes in his masterful book *A Whale Hunt*, the hunt for that whale—being the first hunt that the tribe had carried out in seventy years—was both incredibly difficult for the tribe itself and extremely controversial among anti-whaling advocates and other environmentalists.

Traditionally, the whale hunt in Inupiat communities like Barrow was as much a religious practice as a search for food. The National Park Service brochure for the remarkable Inupiat Heritage Center in Barrow puts it this way: "A successful whaling crew believed that the 60-ton animal had given itself to them as a result of their virtuousness in the preceding year and their rigid adherence to the proper rituals. Inupiaq whaling was both a means of subsistence and a religious ritual, and the two could not be separated." Although whaling in Barrow today may not be quite the religious ritual that it was in the past, the practice continues to be deeply imbued with religious significance, including the continued belief that the whale gives itself to the hunters. Whether whaling in Barrow should technically be considered a religious practice, however, is not ultimately all that important. I made the trip because whether or not whaling "counts" as religion, for the Inupiat community, whaling certainly plays a similar role that religion plays in other small, close-knit communities. As a young whalers' teaching guide that I came across in the town library says, "Whaling is fundamental to our lives. It defines who we are." Studying the conflict between environmentalism and a deep-seated cultural practice like Inupiat whaling, then, can provide relevant lessons for the religious context as well.

More specifically, I figured that learning about whaling and its importance for the Inupiat community might give me a better understanding of the controversy over taking bald and golden eagles in Wyoming and elsewhere. The Native American tribes in the Lower Forty-Eight who want legal permission to take eagles, after all, are seeking something quite analogous to what the native communities in Alaska already have for whales. The trips that I had made before I visited Barrow, which was my last trip for the book, had taught me a lot about how we might manage conflicts between religious practice and the environment, but I was still feeling unsure about the eagle situation. I traveled to Barrow, in other words, to see what the whales could teach me about eagles.

Flying to Alaska from the East Coast takes so long that when you get off the plane, it's a surprise that people are speaking English and using US currency. My original plan had been to spend a couple of days in Anchorage and then maybe four in Barrow. The exact days of the Nalukataq festival aren't set until a week or two before they occur, however. When it turned out that I had guessed wrong about when the festival was going to start (I'd had to buy the tickets a month earlier), I had to change my plans at the last minute so that I could at least catch most of one of the days. As a result, I ended up spending less than a day in Anchorage and over six in Barrow, which, as you will see, is a long time to spend in Barrow.

Making the best of my half-day in Anchorage, I met up with a former student named Jackie, who years ago had left a big Manhattan firm to practice law in Alaska and sing in her own jazz band. Jackie had been to Barrow a couple of times for work and suggested that we go to a grocery store so I could buy some food for my stay. She explained that since everything has to be flown or shipped into Barrow, the prices are far more expensive and the selection far less plentiful then in Anchorage. Jackie advised me specifically to get some produce. "It'll cost you five dollars for an apple in Barrow," she told me. "And it will be an old apple." I filled my basket with bananas, nectarines, nuts, and granola. Jackie suggested I buy an avocado, but I was resistant, because buying an avocado is always like playing the lottery—you never know whether it's going to be completely gross inside. (A Nobel Prize for whoever figures out how to send a tiny person into the middle of an avocado to report back on its condition.) Jackie insisted. "You'll eat it anyway," she said, "because you'll be in Barrow."

Jackie also accompanied me to a liquor store, where I bought a 750-milliliter bottle of Bushmills Irish whiskey for the trip. For many years now, Barrow has been a damp town, meaning that although it is not illegal to drink alcohol there, it is illegal to sell it. The citizens passed the law because alcohol was wreaking havoc in the town, raising the incidence of domestic violence and other crimes, although counterproductively, the law does allow most people to import a certain amount of liquor from out of town every month. Moreover, although it is generally illegal to bring alcohol into Barrow, an exception to the law allows visitors to bring with them up to a liter of spirits, so that's what I decided to do, even though I had an inkling that the hotel where I was staying had a strict policy banning alcohol completely from its premises. The idea of being completely alcohol-free for six days above the Arctic Circle, where the sun never sets, was just too much for me to bear.

The ninety-minute flight from Anchorage to Barrow is more expensive than the flight from Boston to Anchorage, and although the plane was full-sized, at least half of it was blocked off and used for transporting cargo. The ten rows that were reserved for passengers were filled with people who (other than me) clearly all knew each other.

The Barrow airport resembles a small-town bus station. It is too small to fit all the people who use it, and the luggage just gets thrown onto a slanted platform where people compete to grab their stuff first. I took a six-dollar cab ride to the Top of the World Hotel and checked in. The place had just recently been renovated and was quite lovely, at least on the inside. In the lobby, a glass case held a full-sized mastodon tusk underneath a flat-screen, high-definition television. There were indeed prominent signs declaring the hotel's no-alcohol policy and announcing that anyone caught with alcohol would be told to leave, so when I got to my room and unpacked, I wrapped the whiskey bottle in some clothes and buried it deep in my suitcase, where I hoped nobody would come looking for it. There are only three hotels in the town, and the other two were fully booked, so if my hotel kicked me out, I would surely have frozen to death in the parking lot, at least if a polar bear didn't get me first.

A word about polar bears. Everyone in Barrow talks about polar bears, and I was assured by many people that they can be found there, usually out on the ice where they hunt for seals, but also occasionally in the town itself. During my stay, I heard at least three different versions of the "you don't have to be fast enough to outrun the polar bear, you just have to be able to outrun the slowest person in your group" joke. The *Wikitravel* page for Barrow says, "If you do see a Polar Bear stay at least 100 yards away, and preferably in a vehicle. Polar Bears absolutely will eat you." It was probably for the best, then, that my trip ended up being totally, completely, 100 percent polar-bear-free.

After settling in at the hotel, I took the first of many lonely walks that I would take around town throughout the week. Barrow is unlike any other place I have ever visited. It is almost otherworldly, somewhat postapocalyptic—what I imagine it might be like to live on the moon. The land and sea stretch out flat as far as the eye can see—there isn't a tree within hundreds of miles of the place—and the wind whips through town with an intensity that makes it feel cold even when the temperature is above freezing. Since the area receives less than a half foot of precipitation every year (measured in rainfall inches), it is technically a desert, and you can feel the dryness of the air on your skin and lips. The sky is eerie and gray, as the light from the sun, though it never completely disappears during the summer, mostly remains low on the horizon (I can't imagine what the place must be like between November and February, when the sun is never seen at all). The town is quiet, the empty air occasionally punctuated by the sound of a dog barking or an ATV growling down one of the muddy, unpaved roads that crisscross the village. Because the scant snow nevertheless remains on the ground for most of the year, the ramshackle homes are ashen and weather-beaten—the whole place looks like it could use a power wash and several coats of fresh paint. Homes are built on stilts so they don't sink into the ground when the permafrost melts each spring, and the yards are filled with wrecked equipment, broken-down snowmobiles, and all sorts of other bric-a-brac that makes it look as if everyone were holding a yard sale that no customers want to visit.

The outward appearance of the town, however, masks how much Barrow and its residents have going for them. Somewhere between 4,000 and 4,500 people call Barrow home. About 60 percent of the population is Native Alaskan; the others are a mixture of mostly whites and Asians. Some of these non-Natives have come to teach, provide medical services, or do scientific research (the town's location makes it a prime spot to learn about the effects of global warming). The ground nearby is filled with oil, and a lot of money has been invested in the town's public buildings. There's a brand-new state-of-the-art hospital, an eighty-million-dollar high school, and a blue turf football field where the high school Whalers play their home games three miles from the northernmost point in the United States.

My intention was to spend some time talking to people involved with whaling so I could understand the role that it plays in the community. But since my efforts to contact whaling captains had been met with silence, I decided just to try to meet people around town and talk with them. This, however, turned out to be far more difficult than I had thought, largely because there are no public gathering places in Barrow—no shopping malls, no cafés, and certainly no bars. The town does have half a dozen or so restaurants scattered about (e.g., Arctic Pizza and Northern Lights), but other than the one in the hotel, which caters mostly to tourists, they were basically empty when I ate at them. I had never really understood the idea of a pub as a public house before, but in Barrow, I felt the absence of such a place fairly strongly. It certainly made for a week with a lot of empty time, which I filled by watching World Cup games and twenty-two episodes of *Orange Is the New Black*. Desperate for things to do, I watched some of a co-ed softball tournament one afternoon at the dusty dirt "field" behind the public gymnasium and attended a folk music show at the library. The show was put on by a quirky singer named Sunrise, who turned out to be the pop-star Jewel's aunt and who insisted on shaking my right hand with her left hand and vice versa because doing so "makes a circle." I snuck occasional healthy slugs from my contraband bottle of forbidden whiskey.

Of course, it wasn't *all* fun and games during my week in Barrow. Though it was harder than I'd hoped, I did manage to talk with a bunch of people around the town. I chatted with a librarian, the people at the Inupiat Heritage Center, and some of the members of the hotel staff (Cristina, who had just returned from living in Mexico and whose brother is on a whaling crew, was particularly friendly). I took a few tours of the area—two with whaling crew members (Will and Bob) and one with a guy named Mike Shults, whose mother, Fran Tate, famously (she was on The Johnny Carson Show once) owned Pepe's, a Mexican restaurant that was a Barrow icon until it sadly burned to the ground in 2013. I visited Mike's brother Joe's "museum" in his cramped home, a crazy but fascinating hodge-podge of Barrow-area relics and where mastodon bones and baleen sleds share shelf space with a collection of old Big Mac cartons and a Cleveland Cavaliers bobblehead. While I was on Mike's tour, we ran into a whaler who described to us the exact location on the whale where you want to strike it so the bomb (more on this in a bit) goes off in just the right place to kill the animal cleanly.

In the course of these and other conversations, I learned quite a bit about the communal and religious aspects of whaling in Barrow. But first, a few facts:

- Bowhead whales are one of the largest species of whales in the world, growing up to sixty feet or more in length. The Inupiat say that the whales weigh roughly a ton per foot. According to IWC estimates, the worldwide bowhead population is somewhere between eight thousand and twenty-five thousand whales.

- Bowheads are thought to be able to live for up to two hundred years. On one occasion, whalers in Barrow found a stone harpoon tip inside a whale that they had killed, indicating that the whale had been hunted by somebody else before modern whaling techniques had even been invented. According to the whaler who told me this story, the find made him feel united with his ancestors, as though they had both hunted the same exact whale, over a hundred years apart.
- In Barrow, there is both a spring and a fall whaling season. In the spring, the whaling crews hunt more-or-less traditionally, cutting a path through the ice, camping for weeks near the water, and hunting in eight-person wooden boats that are covered with seal skins. When a whale is killed, sixty people are needed to pull it onto the ice, where it is butchered. In the fall, the crews use aluminum motorboats, and bulldozers bring the whales to where they will be processed. People in Barrow don't like to talk very much (at least with outsiders) about the fall whaling season.
- There are forty-plus whaling crews in Barrow. Each crew has a captain, who is responsible for the equipment and safety of the crew. The crew consists of not only the eight people who go in the boat, but also the wives of the crewmembers and other helpers, sometimes young people, who take care of the camp, make the coffee, and the like. Many of the crews have their own distinctive jackets, some with Bible quotes about whales on them. The women who are part of the crew are in charge of distributing the whale meat once the whale is butchered. Traditionally, all of the actual hunters have been men, but during the Nalukataq celebration, one of the captains gave a shout-out to a young woman who had trained on their boat, saying, "Our culture is moving forward."
- Every year, the Alaska Eskimo Whaling Commission allocates Barrow between twenty and twenty-five *strikes*.

Every time a harpoon hits a whale, it counts as a strike. The harpoon is not what kills the whale, however. The harpoon is connected to a darting gun, which shoots an exploding projectile bomb deep into the body of the whale. If the bomb goes in at the right place, it could by itself kill the whale. If not, the hunters carry special shoulder guns, which will finish the job.

- Once the whale is butchered, clear and firm rules govern how it is distributed. Certain parts go to the captain. Others go to the crew who caught the whale. Half of the baleen goes to other crews who helped tow the whale back to shore. One part goes to the harpooner. Some is shared by all whaling crews in town. Still others are divided up and distributed at the Nalukataq or other feasts like those at Christmas and Thanksgiving. It's complicated.

- Families keep their whale meat and other subsistence meat (seal, walrus, caribou, fowl, etc.) in ice cellars that are dug into their yards. Groceries in Barrow are, as Jackie told me in Anchorage, indeed ridiculously expensive. A favorite pastime of visitors to Barrow involves going to the supermarket and looking at the prices and gasping. My favorite item was a twelve-pack of paper towels that was selling for $37.75. A big (but not too big) jar of mayonnaise was retailing at $14.95. Because of these crazy prices, the citizens of Barrow do in fact rely on their subsistence catches for a good deal of their food. People I talked to said they eat whale something like twice a week. Under federal law, the subsistence meat cannot be sold.

- Around town, there are a number of Dumpsters painted with pictures and inspirational messages. My favorite has a picture of a whale and says SAVE THE WHALES . . . FOR DINNER.

It was clear from everything I saw and heard in Barrow that although the purpose of catching the whale is to provide food for the community, the hunt is thoroughly imbued with religious significance. I asked the two whaler tour guides if whaling is a spiritual activity, and both enthusiastically affirmed that it is. The whaling exhibit at the Inupiat Heritage Center, which is where the whaling community expresses itself most clearly to the outside world, also speaks in religious terms about the hunt. A plaque titled "The Whale Gives Itself to Us," for instance, says: "We must be solemn when the whale is dying. We request silence when we strike, and no expression of joy is made until the animal is dead, in respect to the whale's spirit. After the whale is dead, the men surround the whale and say a prayer of thanks for the gift." Another plaque is devoted entirely to the spirit of the whale:

> Our traditions speak of the spirit of the whale as being a flame of light burning in an oil lamp attended by a young girl. The young girl tends to the flame and only steps away to breathe when the great whale surfaces. If the flame were extinguished, the spirit girl would die in the same instant. While men may hunt the physical whale, the spirit of the whale gives itself to the women. The women maintain the sanctity of the home, feed the needy, and care for others. The whale's spirit tells other whales of the kind treatment it received and convinces them to give themselves to the men the following year.

Beyond the religious accoutrements of the whale hunt, however, is the more important point that whaling holds the Inupiat community in Barrow together and gives it meaning. In this way, whaling serves the same function that more obviously religious traditions serve in other communities. Whaling is everywhere in Barrow. It runs through every aspect of the town, from the name of the high school sports teams to the town seal to the food that people eat twice a week for dinner. For a couple of years in the late 1970s, the IWC actually banned subsistence whaling in the Inupiat communities. When I asked the whalers how the ban had affected the town, one man told me that it was "dismal," while the other said that in those couple of years, the community had almost lost hope.

Both men told me that nearly every member of the Inupiat community in Barrow is involved in whaling in some way.

A Japanese researcher who spent many years studying the role of whaling in Barrow and wrote an excellent paper about it summarizes the role of whaling nicely:

> In sum, the whaling activities and feasts are culturally, socially, spiritually, politically, and nutritionally important in contemporary Inupiat society. They also form a basis for their ethnic and community identities. . . . [W]haling and associated feasts of the Inupiat are inseparably related to their contemporary way of life. Thus, the whaling tradition of the Inupiat is fundamental for the cultural and social continuation of the Inupiat as a people.

Perhaps there is no better testimony to the importance of whaling to the community than how young the children start taking part in it. As one of the whalers told me when I asked him about his son, "I want him to start being involved when he turns nine. So it gets into his heart."

Following my conversation with the cackling octogenarian, I had located a good-enough place to sit down on the northern side of the Nalukataq grounds and waited for the festivities to begin. They started with one of the crew captains making some opening remarks in Inupiat and then a long prayer, also in Inupiat, except for the very end, which concluded, "In the name of Jesus, Amen." Most natives who live in Barrow these days practice as Presbyterians, although my guess is that it's with an Inupiat twist. The prayer took place with the members of the two crews—maybe thirty people in all—standing around in a circle and holding hands. Later I would ask somebody what they had been praying about, and the answer was that they were thanking the whale and praying for it, as well as giving thanks to a Christian God for the food.

I knew that the whaling crews were going to serve food to the rest of the people attending the Nalukataq festival, but I had assumed that the guests would get up from where they were sitting and go to the food at some central location rather than the crews bringing the food to us, which is how it actually worked. The soups came first, mostly caribou but also goose and duck. A nice person from one of the crews scooped me out a ladle of caribou soup into my tiny Styrofoam cup. The soup was mostly broth with just a little chunk or two of caribou, but it was delicious and welcome, given the cool weather.

And then came the day's first taste of whale. Over the course of the week, I had tried whale in the form of muktuk several times, provided by the natives who were running the various tours I'd taken. Muktuk is a cut of whale that includes both the blubber and the skin of the animal, usually boiled, sometimes salted, and served in small chunks or strips. It can be chewy, and it's hard to explain what it tastes like—definitely fishy, a little like tuna, with maybe a touch of that indescribable savory umami sensation that the scientists say is the fifth taste—but I had really enjoyed it, and so despite feeling a little guilty about eating this treasured creature, I was nonetheless looking forward to eating more.

The first preparation of whale that the crews served at the feast, however, was a far cry from the relatively benign muktuk. Scooped by hand out of deep buckets of reddish-brown goop, this was what the Inupiats call *mikiaq* or *mikyuk*, which is a mixture of whale meat, skin, blubber, and I think organs, all fermented for a few weeks in the whale's own blood. Mmmm! Having lived in China before, I've eaten a lot of weird things, including dog, snake, and little live shrimp still jumping around in a tangy liquor sauce, but I was not at all thrilled about trying this stuff. I probably wouldn't have, either, except that a guy sitting next to me told me that I should definitely try it. Since he was a member of a whaling crew himself, I felt it would have been rude to object, so I asked one of the bucket people for a "very small" piece, which she happily deposited into my Styrofoam clamshell box.

The whale was surprisingly tough, however, and my little plastic silverware was unequal to the task of cutting it. I was about to give up, when luckily (I guess) the guy next to me saw me struggling and lent me his long, sharp knife, which I used to cut the piece up, although the action also totally shredded the clamshell, rendering that thing thereafter useless. I returned the knife and ate up the *mikiaq*, hoping that it wouldn't kill me. Unfortunately, I don't really have the words to describe how the stuff tasted, but suffice to say, I did not ask for seconds.

During the break that followed, I wandered back to the hotel to warm up a bit and get a new caribou-soup-free coffee cup. When I returned, little kids were taking turns on the blanket toss, and a small group of native singers were singing Christian songs with verses like "Oh God my savior Lord to me" and "One, two, three, devil's after me . . . seven, eight, nine, he missed me all the time." One of the captains came to the microphone to announce that the slab of whale that had been sitting in the middle of the grounds was now cut up and available for visitors to take home with them. I joined a crowd of people—mostly members of other Arctic communities— by the huge blue tub and cardboard boxes that were now filled with large pieces and chunks of whale meat and watched as people jockeyed for space and selected the choicest pieces to put in their Ziploc bags for the trip home. A fellow visitor from New York whom I had met earlier in the morning told me that she thought this was "their Black Friday," which I didn't understand until she explained that the throng of whale-slice choosers looked like mall shoppers battling for the last discounted flat-screen television on the day after Thanksgiving. I stood and watched the free-for-all for a good while and even considered taking a piece myself, which maybe I would have if I'd had any idea what to do with it. As it was, I would have had to put the slice in my coat pocket for the fifteen-hour trip back to Boston.

The next part of the festival day—and the last part for me, because I had to leave at around 5:30 p.m. to catch my plane—involved mass distribution of lots of different parts of the two whales. It was during this time that I finally succeeded in my quest to find someone I could hang out with and talk to about the festival. My earlier efforts on this front had gone nowhere. First, the old lady had laughed at me. Then, a woman named Fluffy showed me to an empty chair that I thought was one of hers, but that actually belonged to someone else who had left for a while and then come back and sent me packing. Finally, a youngish guy I approached to ask about the meaning of the prayer had no idea what it was about but was really interested in asking *me* questions and in fact periodically approached me throughout the afternoon to ask me things about Boston like, "Did you ever see Larry Bird?" and "How'd you like *Good Will Hunting*?"

I had almost given up when I happened to meet a great young guy named Gabe—a native Inupiat who was born in Barrow. Gabe had left Barrow for a while to go to school in Florida before returning to play a kind of liaison role between the natives of Barrow and the outside world in the media and elsewhere. This guy was really funny and knowledgeable, and it was a pleasure to listen to him describe the various roles he's played in Barrow. He has served on a whaling crew, hunted caribou, watched out for polar bears, and done many other things you would expect a young Inupiat guy to do growing up. Gabe explained to me that he *loves* to eat whale, which he claimed is incredibly nutritious in addition to being delicious. He used to carry bags of muktuk in his pockets to munch on during school, and he even tried for a time to go on an all-muktuk diet that he was hoping to write about for a men's health magazine (even he couldn't pull that off for long enough). I asked him whether there was a lot of prayer on the hunt, and he said that there was—before the hunt, during, and after, much of it to Jesus. "This," he said, pointing at the crowd collecting and munching on whale, "is our bread and our wine. Only with more protein."

Over the next couple of hours, the two crews came around to distribute all sorts of different cuts of whale to the community. Gabe collected as many pieces as he possibly could, even making sure that I got my allotted share, which I then graciously donated to him. At the same time, though, he also took small pieces of each cut, sliced them up with an *ulu*, and let me try all of them. There was something called *quaq*, deep-red chunks of whale meat, several different cuts of muktuk—one with a pinkish layer of blubber under the thick, dark layer of skin and another one with more orangey blubber—and a cut from the tail, the name of which I forget, but which was much tougher and less palatable than the other cuts (a woman nearby, when asked why she wasn't eating the tail piece, responded, "I'm done teething"). I don't see myself going out of my way to eat any more whale during my remaining years on the planet, but I do have to say that most of it was pretty delicious.

Witnessing this part of the festival, I had two overwhelming impressions. The first was that *whales are freaking huge!* The amount of whale that the two crews distributed to the people to take home with them in the coolers was amazing. The chunks and pieces and slabs just kept coming and coming. There would be one distribution of the quaq, say, where every person got five or so blocks of meat, and then since there was still plenty of quaq left, the crew would hand out another three blocks to everyone. And then there was the next cut and the next cut, and so on. By the end of the day, there were hundreds of coolers filled to the brim with pieces of whale. Moreover, this was just the first day of the festival. And only a few parts of the whale are distributed at this festival. And this was just the festival for the spring hunt, with the fall hunt still on the way. The whales, in other words, really do continue to feed the town. It's incredible.

The second major impression I had was that *this is what I'd been hearing about all week*. Everything that the Inupiats had told me about the whales giving the community meaning and holding it together and being a source of joy was vividly on display. Everyone was laughing and smiling and joking with each other and talking to friends and family and basically filled with happiness and cheer and hope, and it was all because they are allowed to hunt and kill and eat these whales. When I had first arrived in Barrow, I hadn't been sure what I thought about the practice of hunting whales or whether a subsistence-whaling exception to the global ban was necessary. After my week in Barrow, my thinking on the issue had evolved. Of course, I sincerely hope that those who hunt the whales don't hunt so many of them that the animals become endangered. But sitting there on that cold, sunny day, hundreds of miles north of the Arctic Circle, watching a community bond over an activity that it has practiced for centuries, I believe I understood why the Inupiat need to hunt the bowheads, and I decided that the law was right to let them do it.

When thinking and talking about religion, we have to keep in mind that religious practices are important not only for individuals, but also for communities. Pretty much all of the religious practices described in this book, for instance, have in them a communal aspect that is at least as important as, if not more important than, their individual aspect. Think of Catholics in New York or Boston coming together to celebrate Palm Sunday Mass or the millions of Chinese people in Hong Kong making a pilgrimage to the Yuen Yuen Institute to burn joss paper in its enormous furnace. What about the hundreds of Buddhists outside Taipei congregating at Hai Tao's hilltop temple to worship together and (usually) to release animals into the environment? Or the many millions crowding the streets and beaches of Mumbai to say good-bye for the year to the great Lord Ganesh. Any regulation that the government enacts to protect the environment from these practices, then, will affect not only the many individuals who are practicing their religion, but also the larger religious community. And so, the government, when deciding whether and how to regulate religious practices to protect the environment, must consider the effects of the regulation not only

on individual religious expression but also on the religious community comprising those individuals.

Sometimes, the community-focused nature of a religious practice can reduce the effects of that practice on the environment. One
big bonfire for the entire group might have a smaller environmental
impact, for example, than if each individual started his or her own
fire. If the community can come together over a sacrifice of a single
animal, rather than each person sacrificing a different animal, all
the better. In the case of the whales in Barrow, it's not as if every
person needs to go out and catch a whale. A few catches should do
the trick. Indeed, if the community does not need to eat all twenty
or so whales that it catches every year, it should be limited to catching fewer.

Throughout this book, I've tried to limit my recommendations to
the governments or environmental advocacy groups seeking regulation of a religious practice, and I have not made recommendations aimed at religious individuals or communities themselves. As
I explained in the introduction, I have avoided advising religious
groups, because, as an outsider to those communities, I lack the
authority to suggest that they do or don't do anything. Still, if I had
to plead for religious groups to do one thing differently when it
comes to religious practices that harm the environment, it would
be to think about whether the needs of the community could be
satisfied by collectivizing the religious practice in question. Might
the Hindus in Bangladesh get away with "sacrificing" one turtle instead of a hundred thousand during the Kali Puja festival? Rather
than immersing thousands, could saying thank you and farewell
to one giant Ganesh statue in the sea next to Mumbai suffice? My
hope is that as the religious and environmental communities of the
world continue their discourse on how to keep both sides of the
religion-environment divide happy, collectivizing eco-unfriendly

practices is one possibility that religious communities will agree to take seriously.

Some communities, of course, already do. The Native American tribes that seek whole bald and golden eagles, as well as their feathers, for their religious ceremonies, are not seeking to take an inordinate number of eagles. This restraint should not be surprising, as Native Americans are, on the whole, quite protective of the environment. Indeed, in most controversies involving Native Americans, the government, and the environment, it is the government that is trying to do something harmful to the environment (see, for example, the classic Supreme Court case of *Lyng v. Northwest Indian Cemetery Protective Association*, in which a Native American tribe sued, unsuccessfully, to stop the US Forest Service from building a road right through the tribe's sacred forest).

The potential harm to the environment from the Native American use of eagles and eagle feathers, particularly now that the bald eagle is no longer listed on the endangered species list, is minimal. On the other hand, the eagles play a critical role in preserving the strength and identities of the Native American communities that need them. How the Inupiats of Barrow describe the role of the whale for them is remarkably similar to the Arapaho Tribe's description of the eagle's importance, in an earlier-discussed court case:

> Eagles, in the Arapaho culture, are revered and used for religious ceremonial purposes. . . . An Arapaho does not go out with the purpose to "hunt" an eagle. An eagle presents itself and donates its "holy body" to an Arapaho who needs it for ceremonial purposes such as with the Sun Dance. The Sun Dance is vital to the religion of the Arapaho people. Our deep connection to eagles is a vital and necessary component for the cultural survival and religious identity of the Arapaho people.

Should the Native American tribes that want to take a small number of eagles have the right to take them? This was basically the question that got me started on the journey that has become this book. Well, after thinking about it for several years, and logging a whole lot of miles flying around the world to help me think about it better, I think I've finally found my answer:

Yes! (Probably.)

Okay, so I guess I hedged on that answer a little bit. But I hope you understand by now why I wanted to hedge on it: to emphasize that these questions about how to balance environmental protection and religious freedom are usually hard, quite hard. I'm sure that plenty of people out there will disagree with me on this point and think that the issues are easy. Some nonbelievers will no doubt believe that society, and particularly the law, should not make special accommodations for religious practice. Why should we let people burn bonfires or pollute the water or kill animals for religious reasons, when we wouldn't let them do the same things for nonreligious reasons? On the other hand, some believers will undoubtedly think the questions are easy but reach the exact opposite conclusions. Surely in the big scheme of things, the amount of pollution contributed by religion is so small that society, and particularly the law, shouldn't bother itself with regulating such a fundamental part of so many people's lives just to make a point about equality.

My position is that both of these views are equally wrong and that in most instances when religious practice and environmentalism collide, the law and the rest of society should seek an appropriate balance, one that will differ depending on the context but that will always benefit from a consideration of the lessons I've laid out in these chapters. Religious practices are fundamentally important to those who partake in them, so even those who do not engage in these observances should afford the practices some respect. On the other hand, the natural world is necessary for all of us, and even small harms can have large impacts on the environment and public health.

The key in every case is to take both sides of the equation seriously and to think creatively about how best to protect our precious environment while still allowing religious believers a wide swath of freedom to continue practicing their cherished beliefs. Of course, I am not saying that society will achieve an adequate solution every time—the issues are far too messy to hope that religious believers and environmentalists will be able to compromise and cooperate every time they find themselves at odds. What I am saying, however, is that when it comes to religious practices that harm the environment, the stakes on both sides are simply too high for us not to try.

ACKNOWLEDGMENTS

I would like to thank the following people for their valuable help as I researched, wrote, and published this book: Bernadette Atencio, Helene Atwan, Stephen Awoyemi, Jack Beermann, Amanda Beiner, Amy Bond, Patty Boyd, Chantal Line Carpentier, José Román Carrera, S. Y. Chu, Amy Coleman, Beth Collins, Lorri Cramer, Steven Crook, Dean Current, Shrikant Deodhar, Luis Corzo Dominguez, Kin Fai-Ho, Don Fehr, José Arbey Gomez Garcia, Megan Greene, Tom Hallock, Iris Ho, Tong Chee Kiong, Lin Kuo Chang, Elaine Yi-Jung Lin, Danny Lo, Susan Lumenello, Pam MacColl, John Masland, Carlos Maycotte, Caitlin Meyer, Laison Corzo Montejo, John Nagle, Melissa Nasson, Joe Orifici, Maureen O'Rourke, César A. Cruz Parra, Ali Putnam, S. M. A. Rashid, Jackie Schafer, Li Mau Sheng, Benkong Shi, Joe Shults, Mike Shults, Jorge Sosa, Nancy Tan, Romain Taravella, Gabe Tegoseak, Charlotte Tokos, Jim Tokos, Karen Tokos, Celeste Trujillo, Juan Trujillo, Chu Tseng-Hung (Wu Hung), Raina Isabel Valenzuela, and all the other people in the palm trade in both Guatemala and Chiapas who were nice enough to spend some time with me, as well as Alexser Vázquez Vázquez, Rosa Maria Vidal, Tong Wai Hop, David Walker, Fred Wexler, Mary Wexler, Walter Wexler, Sam Wilgus, Dennis Wilst, Duncan Wong, and Bob, Cristina, and Will from Barrow; Oscar and Lubenay from Pronatura; Vinita, Neel, and Ramanand in Mumbai; and Jen and York in Taipei.

NOTES

INTRODUCTION

Lynn White's lecture can be found in Lynn White Jr., "The Historical Roots of Our Ecological Crisis," *Science*, March 1967, 1203–7. On the Alabama official's claim about EPA's coal regulation, see Stan Diel, "Pray God Blocks EPA Plan, Chief Regulator of Alabama Utilities Tells Consumers," Alabama Media, July 28, 2014, http://www.al.com/news/index.ssf/2014/07/post_14 .html. On the Resisting the Green Dragon series, see Cornwall Alliance for the Stewardship of Creation, "Resisting the Green Dragon," accessed April 24, 2015, http://www.resistingthegreendragon.com/. For a summary of Mike Huckabee's views on religion and the environment, see "Mike Huckabee on Environment," *On the Issues*, last updated February 8, 2010, http:// www.ontheissues.org/2008/Mike_Huckabee_Environment.htm. The website for the Alliance of Religions and Conservation is http://www.arcworld .org/. On the role of environmentalist evangelicals, see David Wheeler, "Greening for God: Evangelicals Learn to Love Earth Day," *Atlantic*, April 18, 2012. For the pope's encyclical, see Pope Francis, *Laudato Si': Encyclical Letter of the Holy Father Francis on Care for Our Common Home*, May 24, 2015, http://w2.vatican.va/content/dam/francesco/pdf/encyclicals /documents/papa-francesco_20150524_enciclica-laudato-si_en.pdf.

1

RELIGIOUS PRACTICE VERSUS THE ENVIRONMENT

On worldwide deaths from air pollution, see Tarik Jasarevic, Glenn Thomas, and Nada Osseiran, "7 Million Premature Deaths Annually Linked to Air Pollution," *World Health Organization*, March 25, 2014, http://www.who .int/mediacentre/news/releases/2014/air-pollution/en/. The quote about the Grinch trying to steal Lag B'Omer comes from Allison Kaplan Sommer, "The Grinch Who Tried to Steal Lag B'Omer," Haaretz.com (Israel), May 8, 2012. On Lag B'Omer generally, see Ron Friedman, "Lag B'Omer Takes Its Fiery Toll," *Times of Israel*, April 29, 2013; "Israel-Environmentalists Against Hundreds of Thousands Lag B'Omer Bonfires," *Jewish Week*, May 7, 2009; R. Ainbinder et al., "[Lag Ba-Omer Bonfires: Is There Any Association Between the Tradition and Asthma and COPD Exacerbations]" (article in Hebrew) *Harefuah* 144, no. 6 (2005): 386–88. On the possibility of Hanukkah candles being harmful, see Shandi P., "Can Hanukkah Candles Poison You?" *The Whole Life* (blog), December 16, 2011, http://thewholelifeblog.wordpress.com/2011/12/16/can-hanukkah-candles-poison-you/.

Dr. Kin-Fai Ho's coauthored article about the impact of incense on air quality in temples in Hong Kong is B. Wang et al., "Characteristics of Emissions of Air Pollutants from Burning of Incense in Temples, Hong Kong," *Science of the Total Environment* 377 (May 1, 2007): 52–60. Two other pieces on incense in temples are C. K. Ho et al., "Adverse Respiratory and Irritant Health Effects in Temple Workers in Taiwan," *Journal of Toxicology and Environmental Health* 68 (2005): 1465–70; and Hsin Ta Hsueh et al., "Health Risk of Aerosols and Toxic Metals from Incense and Joss Paper Burning," *Environmental Chemistry Letters* 10 (March 2012): 79–87. On air pollution caused by fireworks during Diwali, see Nandita D. Ganguly, "Surface Ozone Pollution During the Festival of Diwali, New Delhi, India," *Earth Science India* 2 (October 2009): 224–29. The second expert on firecrackers during Diwali is quoted in Priyangi Agarwal, "Pollution Level During Diwali Goes Up by 30%," *Times of India*, November 13, 2012. On fireworks in China during the New Year celebration, see Austin Ramzy, "Will Chinese New Year Fireworks Make Beijing's 'Crazy Bad' Air Worse?" *Time*, February 8, 2013. On fireworks and Eid, see "'Have Fun This Eid—but Not with Fireworks': Abu Dhabi Police Issue Warning," *7 Days in Dubai*, July 24, 2014. On regulating fireworks in the United States, see Judy Fahys, "Fireworks Loophole: Critics Decry Pollution Exception," *Salt Lake Tribune*, July 22, 2013.

To learn about the worldwide impacts of indoor air pollution, see World Health Organization, "Household Air Pollution and Health," fact sheet, updated March 2014, www.who.int/mediacentre/factsheets/fs292/en/. On the ritualistic use of mercury generally and Arnold Wendroff's campaign to fight it, see Emily Yehle, "EPA Weighs Threats Posed by Mercury Used in Religious Rituals," *New York Times*, May 18, 2011; Lauryn Schroeder et al., "Ritualistic Use of Mercury Remains a Mystery—but Health Effects Aren't," *Medill Reports Chicago*, March 14, 2013, http://news.medill.northwestern

.edu/chicago/news.aspx?id=219201; Leonora LaPeter and Paul de la Garza, "Mercury in Rituals Raises Alarms," *St. Petersburg Times Online,* January 26, 2004. For a piece written by Wendroff himself, see Arnold P. Wendroff, "Magico-Religious Mercury Use in Caribbean and Latino Communities: Pollution, Persistence, and Politics," *Environmental Practice* 7, no. 2 (June 2005): 87–96. The relevant EPA documents on ritualistic use of mercury are EPA Office of Inspector General, *Public Liaison Report, EPA Is Properly Addressing the Risks of Using Mercury in Rituals,* Report No. 2006-P-00031, August 31, 2006; and EPA, *Task Force on Ritualistic Use of Mercury* (Washington, DC: December 2002), http://www.epa.gov/superfund/community/pdfs/mercury.pdf.

On the loss of wetlands in the 1900s, see Marianne de Nazareth, "Approximately 50% of the World's Wetlands Lost During the 20th Century," Countercurrents.org, October 18, 2012. On losing rain forests, see National Geographic, "Rain Forest Threats," accessed April 24, 2015, http://environment.nationalgeographic.com/environment/habitats/rainforest-threats/. The Supreme Court case referred to in the paragraph introducing the RLUIPA is Employment Division v. Smith, 494 U.S. 872 (1990). The text of the RLUIPA can be found at 42 U.S.C. §§ 2000cc, et seq. On a zoning law that limits the number of vegetables a landowner can grow, see M. S., "Where Growing Too Many Vegetables Is Illegal," *Democracy in America* (blog), *Economist,* October 3, 2010, www.economist.com/blogs/democracyinamerica/2010–10-weird_zoning_laws. The Rocky Mountain Church case is Rocky Mountain Christian Church v. Board of County Commissioners of Boulder, Colorado, 613 F.3d 1229 (10th Cir. 2010). Professor Zale's law review article is Kellen Zale, "God's Green Earth? The Environmental Impacts of Religious Land Use," *64 Maine Law Review* 207 (August 2011): 208–37. On green burials, see Lauren Markoe, "Green Burials Reflect Shift to Care for the Body, Soul and Earth," *Huffington Post,* January 26, 2014. On the shaimos controversy, see Margaret F. Bonafide, "Jewish Artifacts Illegally Dumped in N.J.," *USA Today,* March 14, 2013; Zach Patberg, "Burying of Artifacts Remains Unresolved," *Asbury Park (NJ) Press,* October 15, 2010; Alexander Aciman, "God's Garbage in New Jersey," *Tablet,* April 10, 2013.

On the number of people who embark on a religious pilgrimage each year, and on the water bottles left behind during the hajj, see George Webster, "Holy Cities Face Threat from Polluting Pilgrims," CNN.com, November 9, 2011. On damage to the cave in the Himalayas, see Rebecca Byerly, "Massive Hindu Pilgrimage Melting Sacred Glacier," *National Geographic News,*

March 14, 2012. On the Sri Lanka pilgrimage, see Zahrah Imtiaz, "Daunting Task of Keeping Sri Pada Clean: Severe Lack of Facilities for the Millions of Pilgrims Who Climb the Sacred Peak Has Caused Immense Damage to the Environment," *Ceylon Today*, April 13, 2014. On green pilgrimages, see the website of the Green Pilgrimage Project, http://greenpilgrimage.net, where you can find "Green Pilgrimage Handbook" and "Green Guide to Hajj." Also see Miriam Kresh, "Jerusalem Launches World Conference Project for People of Faith," *Green Prophet*, April 10, 2013, www.greenprophet .com/2013/04/jerusalem-symposium-on-green-pilgrimage-april-21-26/.

On the successes of the Clean Water Act and the point-source/ non-point-source distinction, from someone who did get a rash from contact with Boston's Charles River in the 1970s, see James Salzman, "Why Rivers No Longer Burn: The Clean Water Act Is One of the Great Successes in Environmental Law," *Slate*, December 10, 2012. The line about leeches and sludge worms quoted by Salzman comes from the Federal Water Pollution Control Administration itself. On the Amish and non-point-source farming pollution and the quotes about "guns ablazing" and "federal money," see Amanda Peterka, "Amish Farmers in Chesapeake Bay Find Themselves in EPA's Sights," *New York Times*, October 10, 2011. Other sources on this issue include Sindya N. Bhanoo, "Amish Farming Draws Rare Government Scrutiny," *New York Times*, June 8, 2010; and Brian Winter, "Scientists, Amish to Fight Chesapeake Bay Pollution," *USA Today*, February 2, 2010. On the Swartzentrubers, see Associated Press, "Embattled W. Pa. Amish Sect from Cambria County Moving to Upstate NY," *Wall Street Journal*, November 24, 2012 ("going to hell" quote); and Sean D. Hamill, "Religious Freedom vs. Sanitation Rules," *New York Times*, June 13, 2009 ("enforcing stuff that's against our religion"). The Ohio and Michigan cases are Ohio v. Bontrager, 897 N.E. 2d 244 (Newton Falls Municipal Court 2008) and Beechy v. Central Michigan District Health Dept., 475 F.Supp.2d 671 (E.D. Mich. 2007).

The scientific article on Holi is Joy Joseph Gardner and Deepanjali Lal, "Impact of 'Holi' on the Environment: A Scientific Study," *Archives of Applied Science Research* 4, no. 3 (2012):1403–10. On the Ganges River, see Ritu Sharma, "As India's Rivers Turn Toxic, Religion Plays a Part," UCANews .com, January 9, 2014; Joshua Hammer, "A Prayer for the Ganges," *Smithsonian Magazine*, November 2007; Dean Nelson, "Ganges Hit by Alarming Pollution Levels During Kumbh Mela," *Telegraph* (London), February 24, 2013. The book by the famous doctor and writer is Atul Gawande, *Being Mortal: Medicine and What Matters in the End* (New York: Henry Holt, 2014). For a table reporting the safe level of biological oxygen demand for bathing and the level in the Ganges, see Sankat Mochan Foundation, "Pollution in Ganga & Ganga Action Plan Failures," accessed August 21, 2015, www.sankatmochanfoundationonline.org/PollutionofGanga.html.

On the current rate of species extinction, see the World Wildlife Federation, "How Many Species Are We Losing?," accessed April 24, 2015, http://wwf.panda.org/about_our_earth/Biodiversity/biodiversity/. The case involving the Santeria is Church of the Lukumi Babalu Aye v. Hialeah, 508 U.S. 520 (1993). On Candomblé animal sacrifices, see Nivaldo A. Léo Neto et al., "From Eshu to Obatala: Animals Used in Sacrificial Rituals at Candomblé 'Terreiros' in Brazil," *Journal of Ethnobiology and Ethnomedicine* 5, no. 23 (2009). On the Kali Puja turtle sacrifices in Bangladesh, see "100,000 Turtles Sacrificed in Ritual Slaughter to Celebrate Hindu Ritual," *Daily Mail Reporter* (London), October 27, 2011; "Vast Turtle Slaughter in Bangladesh," WildlifeExtra.com, November 2001. On the Shembe and leopard pelts, see Nkepile Mabuse and Vanessa Ko, "Wild Leopards Threatened by Religious Tradition in Africa," CNN.com, September 16, 2012; "Biologist Enters the Fashion Field in a Bid to Save Wild Leopards," AllAboutWildlife.com, January 12, 2011; "Zulu False Dawn: Shembe Faithful Swap Leopardskin for Faux Fur," TheGuardian.com, February 19, 2014; and "Panthera's Faux Fur Leopard Project," accessed April 24, 2015, http://www.panthera.org/node/1471. On the number of elephants killed in recent years, see Brad Scriber, "100,000 Elephants Killed by Poachers in Just Three Years, Landmark Analysis Finds," *National Geographic News*, August 18, 2014. Bryan Christy's classic piece on the connection between religion, ivory, and the killing of elephants is "Blood Ivory," *National Geographic*, October 2012. On Oliver Payne's attempt to get the Vatican to denounce the killing of elephants for their ivory, see Oliver Payne, "Vatican Responds to National Geographic's Correspondence About Religious Use of Ivory," NationalGeographic.com, January 22, 2013. On the destruction of ivory stocks, see Karl Mathiesen, "Does Destroying Ivory Save Elephants?," TheGuardian.com, February 6, 2014.

❷

GUATEMALAN GREENERY

The magazine article about the yellow-eared parrot is Susan McGrath, "Parrot Conservation Changes a Catholic Tradition," *Audubon*, March–April 2012. For newspaper stories about the EcoPalm project, see Marc Lacey, "U.S. Churches Go 'Green' for Palm Sunday," *New York Times*, April 1, 2007;

Stephanie Reighart, "In York, Church Turns New Leaf for Palm Sunday," *York (PA) Daily Record*, March 31, 2012; Ann Rodgers, "Eco-Palms Bring Sustainability to Christian Holy Day," *Pittsburgh Post-Gazette*, March 28, 2010. My *Slate* piece on the EcoPalm project is Jay Wexler, "Are Your Palm Sunday Palms Bad for the Environment?," *Slate*, March 22, 2013.

❸

INDIAN IDOLS

For some general reading about the Ganesh festival and its environmental impacts, see Sarika Bansal, "Ganesh Chaturthi: India's Toxic Festival," *Guardian* (London), September 22, 2010; Marc Abrahams, "God-Awful Pollution of India's Waters," *Guardian* (London), April 30, 2012; Sudeshna Chowdhury, "Green Dream," *Mid-Day* (India), September 6, 2011. The scientific studies mentioned and quoted in the text are, in order of discussion, Anju Vyas et al., "Heavy Metal Contamination Cause of Idol Immersion Activities in Urban Lake Bhopal, India," *Journal of Applied Scientific Environmental Management* 11, no. 4 (December 2007): 37–39; N. C. Ujjania and Azhar A. Multani, "Impact of Ganesh Idol Immersion Activities on the Water Quality of Tapi River, Surat (Gujarat) India," *Research Journal of Biology* 1, no. 1 (2011): 11–15; M. Reddy et al., "Assessment of the Effects of Municipal Sewage, Immersed Idols and Boating on the Heavy Metal and Other Elemental Pollution of Surface Water of the Eutrophic Hussainsagar Lake (Hyderabad, India)," *Environmental Monitoring Assessment* 184 (2012): 1991–2000. The Bombay High Court decision on Janhit Manch's Public Interest Litigation is Janhit Manch v. State of Maharashtra, PIL Writ Petition No. 1325 of 2003, 2008(5)ALLMR2299 (High Court of Judicature at Bombay, Ordinary Original Civil Jurisdiction), July 22, 2008. For more on the case, see "High Court Asks Maharashtra for Report on Idol Immersion," *India Water Review*, June 8, 2011. For the national government's guidelines, see Central Pollution Control Board, Ministry of Environment & Forests, *Guidelines for Idol Immersion* (Shahdara, Delhi: Central Pollution Control Board, June, 2010), http://cpcb .nic.in/upload/NewItems/NewItem_159_Guideline_for_Idol_Imersion .pdf. See also Maharashtra Pollution Control Board, *Suggested Code of Practice for Environment Friendly Celebration of Ganesh Festival* (Mumbai: Maharashtra Pollution Control Board, 2010), http://mpcb.gov.in/images /guidelinesforimmersion.pdf. The National Green Tribunal's decision in the Gujarat case is Sureshbhai Keshavbhai Waghvankar v. State of Gujarat Ors, Application No. 65/2012 (THC), May 9, 2013.

❹

EAGLES

The quotation about the role of the eagle in Arapaho religion, culture, and community is from "Findings and Conclusions of the Northern Arapaho Tribe Regarding the Right to Take Eagles for Religious Purposes," filed October 14, 2013, as an attachment to the Memorandum in Support of Plaintiffs' Motion for Summary Judgment on Remaining Claims, Northern Arapaho Tribe v. Ashe, Civil No. 11-CV-347-J (D. Wyo). The *Audubon* article is Ted Williams, "Golden Eagles for the Gods," *Audubon*, March–April 2001, 30–39. The Supreme Court cases on religious freedom are the following: Reynolds v. United States, 98 U.S. 145 (1878); Sherbert v. Verner, 374 U.S. 398 (1963); Wisconsin v. Yoder, 406 U.S. 205 (1972); Employment Division v. Smith, 494 U.S. 872 (1990); Burwell v. Hobby Lobby, 573 U.S. __, 134 S. Ct. 2751 (2014). For the case of United States v. Friday, see 525 F.3d 938 (10th Cir. 2008). As of this writing, the most recent printed decision in the ongoing case of Northern Arapaho Tribe v. Ashe is the March 12, 2015, decision issued by federal district court judge Alan B. Johnson, https://cases.justia.com/federal/district-courts/wyoming/wydce/2:2011cv00347/23407/93/0.pdf?ts=1426254665. The decision describes in full detail the history of the litigation. My description of the Iowa Tribe's eagle aviary appeared, in somewhat different form, in Jay Wexler, "Not Your Ordinary Home for Birds," April 11, 2013, *National Geographic News Blog*, http://voices.nationalgeographic.com/2013/04/11/not-your-ordinary-home-for-birds/.

❺

SINGAPORE SMOKE

On the actor's Facebook posting about joss, see "Local Celeb Says Joss Paper Burning Destroys Grass and Causes Global Warming," *Stomp* (blog), *Singapore Press Holdings*, September 1, 2012. For the comment about "dumbasses," see "Spike in Complaints Against Joss Paper Burning," SgForums.com (blog), Singapore, May 8, 2013. For the post comparing joss burning in Singapore to forest fires in Indonesia, see "Burning of Joss Paper," SgForums.com (blog), Singapore, August 22, 2000. For the person without religious tolerance, see "Seriously WTF Is Wrong with These Idiots?," SgForums.com (blog), Singapore, March 1, 2010. For the flowerpot story, see "Man Dies After Quarrelling with Neighbours over Burning Incense," *AsiaOne*

(Singapore), August 15, 2011. Professor Tong's academic article is Tong Chee Kiong and Lily Kong, "Religion and Modernity: Ritual Transformations and the Reconstruction of Space and Time," *Social & Cultural Geography* 1, no.1 (2000). The scientific journal articles about the environmental effects of joss paper burning are Hsi-Hsien Yang et al., "Polycyclic Aromatic Hydrocarbon Emissions from Joss Paper Furnaces," *Atmospheric Environment* 39 (January 2005): 3305–12; Jui-Yeh Rau et al., "Characterization Of Polycyclic Aromatic Hydrocarbon Emissions from Open Burning of Joss Paper," *Atmospheric Environment* 42 (March 2008): 1692–1701; Yu-Yun Lo et al., "Removal of Particulates from Emissions of Joss Paper Furnaces," *Aerosol and Air Quality Research* 11 (2011): 429–36; and M. D. Lin et al., "Characterizing PAH Emission Concentrations in Ambient Air During a Large-Scale Joss Paper Open-Burning Event," *Journal of Hazardous Materials* 156 (2008): 223–29. On the Phoh Kiu Siang T'ng temple, see Daryl Chin, "Eco-Friendly Incense Burners for Temple," *Straits Times* (Singapore), July 18, 2013. On the Kong Meng San Phor Kark See Monastery, see Yen Fang, "Sin Ming Residents Move to Keep a Lid on Temple's Ash," *Straits Times* (Singapore), January 29, 2012; Melody Zaccheus, "Clearer Skies, Roads for Temple's Neighbours," *Straits Times* (Singapore), April 1, 2013. See also Hong Kong Environmental Protection Department, *Guidelines on Air Pollution Control for Joss Paper Burning at Chinese Temples, Crematoria, and Similar Places* (Hong Kong: HKSAR Government, September 2011), http://www.epd.gov.hk/epd/english/compliance_ass/others/files/EPD_Joss_Paper_eng.pdf. For more on the green temple at Wong Tai Sin, see Guo Jiaxue, "Worship Goes Green," *Hong Kong Focus*, March 4, 2011.

6

TAIWANESE TURTLES

To read generally about mercy release and its negative impacts on the environment, see Elizabeth Hsu, "'Release of Life' Religious Practice Spurs Big Business," *Taiwan News*, October 2, 2009; "Groups Release 200 Mil. Animals Annually: Study," *China Post*, October 3, 2009; Amber Wang, "Environmentalists and Buddhists Go Head to Head in Taiwan," *Telegraph* (London), February 22, 2010; Liu Meng, "Snake Bites Dog, Capital Rattled," *Global Times* (China), June 12, 2012; Angelica Oung, "Conference Pans Animal Releases," *Taipei Times*, September 30, 2007. The Humane Society International's web page on mercy release can be found at http://www.hsi

.org/issues/mercy_release/. For the Humane Society International quotations, see Humane Society International, *Mercy Release: Kind Intentions, Cruel Consequences* (Washington, DC: Humane Society International, 2009), which can be found at http://www.hsi.org/assets/pdfs/mercyrelease _flyer_english.pdf. The scientific paper studying the effects of mercy release of American bullfrogs in Yunnan Province, China, is Xuan Liu et al., "The Influence of Traditional Buddhist Wildlife Release on Biological Invasions," *Conservation Letters* 5 (February 2012): 107–14. The RCRC letter is Awoyemi et al., "Mobilizing Religion and Conservation in Asia," *Science* 338 (December 21, 2012). For Benkong Shi's work (but with birds rather than turtles), see Rachel Nuwer, "A Buddhist Ritual Gets an Ecologically Correct Update," *Audubon*, January–February 2014. For more on Wu Hung (Chu Tseng-hung), see Chen Wei-tzu and Jake Chung, "Former Monk Leads Fight for Animal Rights," *Taipei Times*, December 27, 2012. Li Mau Sheng's report for the government is available only in Chinese, but the abstract has been translated into English: Li Mau Sheng, [*Evaluation Program About Legislation of Commercial/Large-Scale Animal Release*] (May 20, 2011). You can watch the touching video of the release of the macaque (but bring tissues) at wfas309, "Wildlife First Aid Station," posted November 11, 2013, www.youtube.com/watch?v=jP-oHJoohoQ. For the Supreme Court declaration that courts cannot inquire into the truth of a claimant's religious beliefs but can inquire into the claimant's sincerity, see Ballard v. United States, 322 U.S. 78 (1944). The case involving the Jehovah's Witness was Thomas v. Review Board, 450 U.S. 707 (1981).

❼

BARROW (ALASKA) BOWHEADS

An enormous amount of information on whaling can be found on the website of the International Whaling Commission, http://iwc.int/home. The masterful book about the Makah's first whale hunt in decades is Robert Sullivan, *A Whale Hunt: How a Native American Village Did What No One Thought It Could* (New York: Scribner, 2000). The young whalers' teaching guide that I found in the Barrow library is North Slope Borough School District, *Whaling Standards Barrow and Wainwright: Honoring the Learning of Our Young Whalers* (Barrow, AK: North Slope Borough School District, 2002). The *Wikitravel* page for Barrow is http://wikitravel .org/en/Barrow_%28Alaska%29. To read about Fran Tate and her Mexican restaurant, see Lyn Kidder, *Tacos on the Tundra: The Story of Pepe's North of the Border* (Anchorage, AK: Bonaparte Books, 1996). On the role of whaling in Barrow, see Nobuhiro Kishigami, "Aboriginal Subsistence Whaling in Barrow, Alaska, Anthropological Studies of Whaling," *SENRI Ethnological Studies* 84 (2013): 101–20.